THE FOURTH DIMENSION

THE FOURTH DIMENSION

BY

E. H. NEVILLE

LATE FELLOW OF TRINITY COLLEGE, CAMBRIDGE
PROFESSOR OF MATHEMATICS IN UNIVERSITY COLLEGE, READING

CAMBRIDGE
AT THE UNIVERSITY PRESS
1921

CAMBRIDGE
UNIVERSITY PRESS

University Printing House, Cambridge CB2 8BS, United Kingdom

Cambridge University Press is part of the University of Cambridge.

It furthers the University's mission by disseminating knowledge in the pursuit of education, learning and research at the highest international levels of excellence.

www.cambridge.org
Information on this title: www.cambridge.org/9781316633328

First published 1921
First paperback edition 2016

A catalogue record for this publication is available from the British Library

ISBN 978-1-316-63332-8 Paperback

NOTE

MY thanks are due to my colleague Mr H. Knapman and my brother Mr B. M. Neville. A long discussion of the manuscript with the latter determined the character of the early paragraphs, and both have read the proofs for me.

CONTENTS

THE FOURTH DIMENSION

INTRODUCTION

To the general reader, the name of the fourth dimension brings reminiscences of *Flatland* and *The Time Machine*. On hearing that to the mathematician the extension from three dimensions to four or five is trivial, he thinks he is being told that a study of mathematics, if reasonably intense, creates physical faculties or powers of visualisation with which the uninitiated are not endowed. Learning that Minkowski and Einstein combine space and time into a single continuum, he tries to believe in the existence of a state of mind in which the *sensations* of space and time are confused, and naturally he fails.

The position of students of mathematical physics, and of all but a fortunate few of the students of pure mathematics, is little better. Accustomed to regard a Cartesian frame of axes as a scaffolding erected in the real space around them, they can attach no *meaning* to a fourth coordinate, but having used complex electromotive forces with success in the theory of alternating currents, and having treated a symbol of differentiation as a detachable algebraic variable even to the extent of resolving operators into partial fractions for the solution of differential equations, these students are prepared to give pragmatical sanction to the most fantastic language.

The pure mathematician makes no attempt to *imagine* a space of four dimensions; he lays no claim to visualising a world that is inconceivable to other men. Only he finds that certain notions in *algebra* are discussed most readily in terms adopted from geometry and given a meaning entirely algebraic, and since it is to the mathematician alone that algebraic problems are of concern in themselves, fear lest the man in the street should mistake the very subject of a mathematical conversation he might overhear has not prevented the mathematician from using the vocabulary he finds best suited to his own needs. Now it has happened that the talk of a few mathematicians has suddenly become of universal and absorbing interest, and a dictionary explaining the meanings they are in the habit of giving to some familiar words is required.

It is this dictionary that I have tried to write, and I have written it in the simplest terms I could find, in the hope that it will prove intelligible to anyone familiar with elementary trigonometry and with the solution of simultaneous linear equations in algebra; for this reason, I have *not* treated the point as indefinable, I have supposed the numbers used always to be real, and I have avoided Frege-Russell definitions. The reader's first feeling will be one of disillusion. Are Einstein and Eddington talking not about a new heaven and a new earth but about linear algebraic equations? To discuss the question is beyond the province of a lexicographer.

Perhaps even the mathematical student, if he can overcome a reasonable irritation at the restrictions, from his point of view arbitrary, to four dimensions and to real numbers, and at the absence of certain obvious forms of abbreviation, may derive some help from the pamphlet. The possibility of constructing an abstract 'space' is always assumed, but the details of the construction, even for two or three dimensions only, are either taken for granted or disguised as theorems on matrices or on linear equations. The idea of direction and the measurement of angles in a constructed plane demand careful consideration. The nature of pure rotation in four dimensions is by no means obvious; on the contrary, rotation is the most difficult of the elementary notions used in the theory of relativity, and with an account of rotation our formal work comes to a natural end.

1. POINTS

1·1. The fundamental idea in the work before us is that of *an ordered group of four numbers*; $(2, -1, 3, 0)$, $(\frac{1}{3}, \frac{2}{5}, -\frac{1}{2}, -\frac{1}{4})$, $(-\sqrt{2}, 1, 1, \sqrt[3]{5})$ are particular groups of the kind contemplated, and (ξ, η, ζ, τ) is a form that can be used to represent any such group, the group becoming definite when actual numerical values are assigned to ξ, η, ζ, τ.

By calling the groups *ordered* we mean that the group (a, b, c, d) is regarded as different from a group such as (a, d, c, b) obtained by interchanging two of its constituent numbers, unless the numbers interchanged happen to be equal.

Each group is to be treated as one composite individual, to be combined or compared with other groups according to rules that are laid down. There is nothing peculiarly mathematical in dealing in this way with a structure that can *if necessary* be analysed, or in having a name for a group as a whole; on the contrary, not only collective nouns such as 'team' and 'nation' but common nouns such as 'book' and 'tree' stand for objects that it is convenient to regard as wholes although it is easy to distinguish parts of which they are formed.

An ordered group of four numbers may be called a *tetrad*. This name evokes no preconceptions and suggests no analogies, and the use of an unfamiliar word emphasises that theorems asserted *must* be founded on definitions given. But the definitions that are natural lead to theorems that *can be made* verbally similar to theorems in geometry by the introduction of terms derived from geometry. The vocabulary of geometry begins with the word 'point', and the mathematician finds it worth while to give to the ordered group of four numbers the *name* of 'point in four-dimensional space'. To him four-dimensional space *is* simply the totality of tetrads under another name. The conception of such a totality is no more abstruse than that of 'all the pairs of numbers whose squares differ by 5'; the name given to it must not be allowed to mislead.

In this pamphlet we shall use the word 'point' simply as an abbreviation for 'point in four-dimensional space', for we are concerned with no other application of the word.

1·2. The four constituent numbers that determine a point are called the *absolute coordinates* of the point. In the examples of points given in the first paragraph, the first absolute coordinate of the first point is 2, the fourth absolute coordinate of the second point is $-\frac{1}{4}$, and the third point has both its second absolute coordinate and its third equal to 1. Often it is convenient to use single letters to denote points and to denote the absolute coordinates of a point P by $\xi_P, \eta_P, \zeta_P, \tau_P$; for example, to write

1·21 $$\xi_A = -\tfrac{7}{3}, \quad \eta_A = \tfrac{1}{2}, \quad \zeta_A = 0, \quad \tau_A = -\tfrac{4}{5}$$

is one way of expressing that A is to be used for a time as a symbol for the particular tetrad $(-\frac{7}{3}, \frac{1}{2}, 0, -\frac{4}{5})$.

There is one point whose absolute coordinates are all zero; this point is called the absolute origin and is denoted by O.

1·3. The symbol of equality is used of points only to express actual identity. The points

$$(a, b, c, d), \ (e, f, g, h)$$

are identical if and only if the four conditions

1·31 $$a = e, \quad b = f, \quad c = g, \quad d = h$$

are all satisfied, and it is therefore natural to use

1·32 $$(a, b, c, d) = (e, f, g, h)$$

as a compact substitute for the set of equations 1·31. If the points are denoted by P, Q, their identity is expressed simply by $P = Q$.

2. STEPS AND VECTORS

2·1. The ordered pair of points (P, Q) is described as the *step* PQ, and is said to have P for its beginning and Q for its end. The absolute coordinates of Q can be derived from the absolute coordinates of P by the addition of the numbers $\xi_Q - \xi_P$, $\eta_Q - \eta_P$, $\zeta_Q - \zeta_P$, $\tau_Q - \tau_P$; these numbers form a tetrad, that is, a point, but our language would acquire an unfamiliar tone if we were to speak of being directed from one point to another point by means of a third *point*, and therefore we avail ourselves for the tetrad in this connection of the alternative name of *vector*, which is introduced for use whenever the name of point is undesirable. There is no difference in definition between the point (a, b, c, d) and the vector (a, b, c, d), but the point is said rather to *represent* the vector than to *be* the vector.

Single letters used to denote tetrads that are being regarded as vectors will be in Clarendon type. The vector $\mathbf{r}$ is the tetrad $(\xi_{\mathbf{r}}, \eta_{\mathbf{r}}, \zeta_{\mathbf{r}}, \tau_{\mathbf{r}})$; the constituent numbers of the tetrad are called the *absolute components* of the vector. If P, Q are points, the vector $(\xi_Q - \xi_P, \eta_Q - \eta_P, \zeta_Q - \zeta_P, \tau_Q - \tau_P)$, which is called the vector of the step PQ, is sometimes denoted by $Q - P$. Given a point P and a vector $\mathbf{r}$, there is one and only one step from P which has the vector $\mathbf{r}$; this is the step from P to $(\xi_P + \xi_{\mathbf{r}}, \eta_P + \eta_{\mathbf{r}}, \zeta_P + \zeta_{\mathbf{r}}, \tau_P + \tau_{\mathbf{r}})$, and the point which is the end of the step is said to represent $\mathbf{r}$ with respect to P. The point whose absolute coordinates are the absolute components of $\mathbf{r}$ represents the vector $\mathbf{r}$ with respect to the absolute origin O.

2·2. Two steps are said to be *congruent* if they have the same vector. The congruence of the steps PQ, RS is expressed symbolically in the form

2·21 $$Q - P = S - R.$$

Since the set of equations

2·22 $$\xi_Q - \xi_P = \xi_S - \xi_R, \quad \eta_Q - \eta_P = \eta_S - \eta_R,$$
$$\zeta_Q - \zeta_P = \zeta_S - \zeta_R, \quad \tau_Q - \tau_P = \tau_S - \tau_R$$

is equivalent to

2·23 $$\xi_R - \xi_P = \xi_S - \xi_Q, \quad \eta_R - \eta_P = \eta_S - \eta_Q,$$
$$\zeta_R - \zeta_P = \zeta_S - \zeta_Q, \quad \tau_R - \tau_P = \tau_S - \tau_Q,$$

congruence of PQ with RS is equivalent to congruence of PR with QS. Moreover, the symbolical expression of the congruence in the latter form is

2·24 $$R - P = S - Q,$$

and comparison of this with 2·21 shews that the straightforward use of the notation will not mislead.

Ex. i*. The pair of points $\{(2, -1, 4, 1), (4, 3, -2, 1)\}$ is a step whose vector is the tetrad $(2, 4, -6, 0)$; the step from the point $(-4, 1, 3, 3)$ with the same vector ends at the point $(-2, 5, -3, 3)$.

2·3. By the product of the vector $\mathbf{r}$ by the number k, positive, zero, or negative, is meant the vector $(k\xi_\mathbf{r}, k\eta_\mathbf{r}, k\zeta_\mathbf{r}, k\tau_\mathbf{r})$; this vector is denoted by $k\mathbf{r}$.

The sum of any finite number of vectors is definable as the vector each of whose absolute components is the sum of the absolute components of the individual vectors. That is to say, the expression

2·31 $$(\xi_1, \eta_1, \zeta_1, \tau_1) + (\xi_2, \eta_2, \zeta_2, \tau_2) + \ldots + (\xi_n, \eta_n, \zeta_n, \tau_n)$$

is defined to denote the vector

2·32
$$(\xi_1 + \xi_2 + \ldots + \xi_n, \eta_1 + \eta_2 + \ldots + \eta_n, \zeta_1 + \zeta_2 + \ldots + \zeta_n, \tau_1 + \tau_2 + \ldots + \tau_n).$$

Since a change in the order of the vectors in 2·31 only deranges the terms in the numerical sums which give the constituents in 2·32, any such change is without effect on the constituents, and therefore is without effect on the vector itself: the sum of a number of vectors does not depend on the order in which the vectors are taken.

If starting from any point Q_0 we form a chain of points

* To be brief and clear, the examples are constructed for the most part with whole numbers; it must be remembered that no restriction of this kind is imposed by the definitions.

$Q_0 Q_1 Q_2 \ldots Q_n$ such that the successive steps $Q_0 Q_1, Q_1 Q_2, \ldots, Q_{n-1} Q_n$ have given vectors $\mathbf{r}_1, \mathbf{r}_2, \ldots, \mathbf{r}_n$, then

$$2{\cdot}33 \quad \xi_{Q_1} - \xi_{Q_0} = \xi_{\mathbf{r}_1}, \ \xi_{Q_2} - \xi_{Q_1} = \xi_{\mathbf{r}_2}, \ \ldots, \ \xi_{Q_n} - \xi_{Q_{n-1}} = \xi_{\mathbf{r}_n},$$

and by addition

$$2{\cdot}34 \qquad \xi_{Q_n} - \xi_{Q_0} = \xi_{\mathbf{r}_1} + \xi_{\mathbf{r}_2} + \ldots + \xi_{\mathbf{r}_n};$$

similar results follow for the other coordinates, and therefore $\mathbf{r}_1 + \mathbf{r}_2 + \ldots + \mathbf{r}_n$ is the vector of the step $Q_0 Q_n$. It is this proposition which gives importance to the vector $\mathbf{r}_1 + \mathbf{r}_2 + \ldots + \mathbf{r}_n$, but to define the sum by means of the chain of steps would render it necessary to prove the sum independent not only of the order in which the vectors were taken but also of the point Q_0 from which the chain was begun. It is to be noticed that the relations between the points and the vectors can be written in the form

$$2{\cdot}35 \qquad Q_1 - Q_0 = \mathbf{r}_1, \ Q_2 - Q_1 = \mathbf{r}_2, \ \ldots, \ Q_n - Q_{n-1} = \mathbf{r}_n,$$

and that the conclusion that would be reached by adding the equations as if the symbols were algebraic is the correct conclusion

$$2{\cdot}36 \qquad Q_n - Q_0 = \mathbf{r}_1 + \mathbf{r}_2 + \ldots + \mathbf{r}_n.$$

Ex. ii. The step from (4, 3, −2, 1) with the vector (0, 1, 4, −2) ends at (4, 4, 2, −1). If we add this vector to the vector of the step {(2, −1, 4, 1), (4, 3, −2, 1)}, we have the vector (2, 5, −2, −2), which is the vector of the step straight from (2, −1, 4, 1) to (4, 4, 2, −1). The same addition can be performed by regarding (2, 4, −6, 0), (0, 1, 4, −2) as the vectors of

{(−4, 1, 3, 3), (−2, 5, −3, 3)}, {(−2, 5, −3, 3), (−2, 6, 1, 1)}

and (2, 5, −2, −2) as the vector of {(−4, 1, 3, 3), (−2, 6, 1, 1)}.

2·4. Subtraction may be defined in terms of addition or by means of the absolute components. The equation

$$2{\cdot}41 \qquad \mathbf{r} - \mathbf{s} = \mathbf{t}$$

is equivalent to

$$2{\cdot}42 \qquad \mathbf{r} = \mathbf{s} + \mathbf{t}$$

and also to the set of equations

$$2{\cdot}43 \quad \xi_{\mathbf{r}} - \xi_{\mathbf{s}} = \xi_{\mathbf{t}}, \ \eta_{\mathbf{r}} - \eta_{\mathbf{s}} = \eta_{\mathbf{t}}, \ \zeta_{\mathbf{r}} - \zeta_{\mathbf{s}} = \zeta_{\mathbf{t}}, \ \tau_{\mathbf{r}} - \tau_{\mathbf{s}} = \tau_{\mathbf{t}}.$$

If $-\mathbf{r}$ is defined to mean the product of $\mathbf{r}$ by -1, which is the vector $(-\xi_{\mathbf{r}}, -\eta_{\mathbf{r}}, -\zeta_{\mathbf{r}}, -\tau_{\mathbf{r}})$, subtraction of $\mathbf{r}$ is equivalent to addition of $-\mathbf{r}$.

If any vector is subtracted from itself, the result is the vector (0, 0, 0, 0), which is called the zero vector. This is the vector of any step in which the end coincides with the beginning. Just as it is often advantageous to express a linear relation

$$p_1x_1 + p_2x_2 + \dots + p_mx_m = q_1y_1 + q_2y_2 + \dots + q_ny_n$$

between algebraic variables in the form

$$p_1x_1 + p_2x_2 + \dots + p_mx_m - q_1y_1 - q_2y_2 - \dots - q_ny_n = 0,$$

so the linear relation

2·44 $$p_1\mathbf{r}_1 + p_2\mathbf{r}_2 + \dots + p_m\mathbf{r}_m = q_1\mathbf{s}_1 + q_2\mathbf{s}_2 + \dots + q_n\mathbf{s}_n$$

between vectors is usefully taken in the form

2·45
$$p_1\mathbf{r}_1 + p_2\mathbf{r}_2 + \dots + p_m\mathbf{r}_m - q_1\mathbf{s}_1 - q_2\mathbf{s}_2 - \dots - q_n\mathbf{s}_n = (0, 0, 0, 0).$$

For brevity, the zero vector (0, 0, 0, 0) is denoted simply by **0**. We have, for any vector **r**,

2·46 $$\mathbf{r} + \mathbf{0} = \mathbf{r}, \quad 0 \times \mathbf{r} = \mathbf{0},$$

and for any number k,

2·47 $$k \times \mathbf{0} = \mathbf{0}.$$

The vector **0** is the vector that is represented by the origin.

A vector that is not the zero vector is said to be a *proper* vector. The equation

2·48 $$k\mathbf{r} = \mathbf{0}$$

implies that either k is zero or **r** is the zero vector.

Ex. iii. There is nothing elusive about a zero step. The step

$$\{(3, \tfrac{1}{2}, 0, -\tfrac{1}{3}), (3, \tfrac{1}{2}, 0, -\tfrac{1}{3})\}$$

is no more subtle than any other group of eight numbers regarded as a pair of tetrads.

3. VECLINES, VECPLANES, AND VECSPACES

3·1. If $\mathbf{r}_1, \mathbf{r}_2, \dots, \mathbf{r}_n$ are given vectors and $x_1, x_2, \dots, x_n$ are unknown numbers, the equation

3·11 $$x_1\mathbf{r}_1 + x_2\mathbf{r}_2 + \dots + x_n\mathbf{r}_n = \mathbf{0}$$

is equivalent to the set of four simultaneous algebraic equations

3·12 $$\begin{cases} x_1\xi_1 + x_2\xi_2 + \dots + x_n\xi_n = 0, \\ x_1\eta_1 + x_2\eta_2 + \dots + x_n\eta_n = 0, \\ x_1\zeta_1 + x_2\zeta_2 + \dots + x_n\zeta_n = 0, \\ x_1\tau_1 + x_2\tau_2 + \dots + x_n\tau_n = 0, \end{cases}$$

where ξ_m, η_m, ζ_m, τ_m are the absolute components of $\mathbf{r}_m$. The equations in this set are of course all satisfied if $x_1, x_2, \ldots, x_n$ are all zero; since only the ratios of $x_1, x_2, \ldots, x_n$ to one another are involved in the equations, we infer from 3·12 that a solution of 3·11 with $x_1, x_2, \ldots, x_n$ not all zero is certainly possible if n is greater than 4 but is not possible if n is not greater than 4 unless the vectors $\mathbf{r}_1, \mathbf{r}_2, \ldots, \mathbf{r}_n$ are related to each other in some special way. We say that the vectors $\mathbf{r}_1, \mathbf{r}_2, \ldots, \mathbf{r}_n$ are *linearly related* if there are numbers $x_1, x_2, \ldots, x_n$ not all zero which render 3·11 an identity; if all the vectors are proper, *two* or more of the numbers must be different from zero. The relation 3·11 is not distinguished from the relation obtained by multiplying throughout by a constant other than zero, and the deduction we make from 3·12 can be expressed in the form that between any five or more vectors there is at least one effective linear relation, the word 'effective' conveying the condition that the coefficients are not all zero.

For the theory of systems of linear algebraic equations, the reader is referred to books on algebra. Perhaps the best account for the English student is in Bôcher's *Introduction to Higher Algebra* (The Macmillan Co., 1907), a masterpiece of exposition.

3·2. Two vectors $\mathbf{a}, \mathbf{b}$ are said to be *collinear* if there are numbers a, b not both zero such that

3·21 $$a\mathbf{a} + b\mathbf{b} = \mathbf{0}.$$

If $\mathbf{a}$ is the zero vector, we can satisfy 3·21 by taking b to be zero, whatever the vector $\mathbf{b}$: the zero vector is collinear with every vector. If $\mathbf{a}$ is not $\mathbf{0}$, then b is not zero, and 3·21 is equivalent to

3·22 $$\mathbf{b} = (-a/b)\,\mathbf{a}:$$

the vectors collinear with a proper vector are the multiples of that vector.

The vectors collinear with a proper vector $\mathbf{a}$ compose a class of vectors that will be called the *vecline** built on $\mathbf{a}$. If $\mathbf{r}$ is any member of this class, there is by hypothesis one number $x_{\mathbf{r}}$ which is such that

3·23 $$\mathbf{r} = x_{\mathbf{r}}\mathbf{a},$$

and this number is unique, for the two equations

$$\mathbf{r} = x_{\mathbf{r}}'\mathbf{a}, \quad \mathbf{r} = x_{\mathbf{r}}''\mathbf{a}$$

* To the best of my belief, this word and others on the same model that will be used later are new.

imply $$(x_{\mathbf{r}}' - x_{\mathbf{r}}'')\,\mathbf{a} = \mathbf{0}$$
and therefore $$x_{\mathbf{r}}' = x_{\mathbf{r}}'',$$
since **a** is proper.

If **f** is a proper vector in the vecline built on **a**, then since every multiple of **f** is a multiple of **a**, every vector that belongs to the vecline built on **f** belongs to the vecline built on **a**; but if **f** is proper and equal to $x_{\mathbf{f}}\mathbf{a}$, the coefficient $x_{\mathbf{f}}$ is not zero and **a** is expressible as $(1/x_{\mathbf{f}})\,\mathbf{f}$; hence **a** is a proper vector in the vecline built on **f**, and every vector that belongs to the vecline built on **a** belongs to the vecline built on **f**. It follows that the vecline built on **f** coincides with the vecline built on **a**, that is, that no two veclines have a proper vector in common; in other words, the vecline built on a proper vector **a** is the only vecline of which **a** is a member, and this vecline can therefore be described simply as the vecline that contains **a**. But every vecline contains the zero vector.

Ex. iv. The vecline containing $(2, 1, -2, -1)$ is the class of all tetrads of the form $(2x, x, -2x, -x)$. The vector $(-3, -\frac{3}{2}, 3, \frac{3}{2})$, which is the product of the original vector by $-\frac{3}{2}$, belongs to the vecline. The vector $(5, \frac{5}{2}, 15, -\frac{5}{2})$ does not, for the *only* multiple of $(2, 1, -2, -1)$ whose first constituent is 5 is the product by $\frac{5}{2}$, and this differs from $(5, \frac{5}{2}, 15, -\frac{5}{2})$ in the third constituent. The vector $x(2, 1, -2, -1)$ is expressible as $-\frac{2}{3}x(-3, -\frac{3}{2}, 3, \frac{3}{2})$, and conversely the vector $X(-3, -\frac{3}{2}, 3, \frac{3}{2})$ is identical with $-\frac{3}{2}X(2, 1, -2, -1)$; that is, every vector in the vecline built on either of the vectors $(2, 1, -2, -1)$, $(-3, -\frac{3}{2}, 3, \frac{3}{2})$ belongs to the vecline built on the other of these two vectors.

3·3. Three vectors **a**, **b**, **c** are said to be *coplanar* if there are numbers a, b, c not all zero such that

3·31 $$a\mathbf{a} + b\mathbf{b} + c\mathbf{c} = \mathbf{0}.$$

If **a** and **b** are collinear, there are values of a and b not both zero such that $a\mathbf{a} + b\mathbf{b}$ is the zero vector, and we can satisfy 3·31 by taking a and b to have such values and c to be zero, whatever the vector **c**: if two vectors are collinear, every vector is coplanar with them. If **a** and **b** are not collinear, 3·31 can not be satisfied if c is zero, and is therefore equivalent to

3·32 $$\mathbf{c} = (-a/c)\,\mathbf{a} + (-b/c)\,\mathbf{b}.$$

The vectors coplanar with two vectors **a**, **b** that are not themselves collinear compose a class that will be called temporarily the *vecplane* built on **a** and **b**. If **r** belongs to this vecplane, there is a pair of numbers $(x_{\mathbf{r}}, y_{\mathbf{r}})$ such that

3·33 $$\mathbf{r} = x_{\mathbf{r}}\mathbf{a} + y_{\mathbf{r}}\mathbf{b},$$

and this pair of numbers is unique, for the two equations

$$\mathbf{r} = x_{\mathbf{r}}'\mathbf{a} + y_{\mathbf{r}}'\mathbf{b}, \quad \mathbf{r} = x_{\mathbf{r}}''\mathbf{a} + y_{\mathbf{r}}''\mathbf{b}$$

imply

$$(x_{\mathbf{r}}' - x_{\mathbf{r}}'')\,\mathbf{a} + (y_{\mathbf{r}}' - y_{\mathbf{r}}'')\,\mathbf{b} = \mathbf{0},$$

and to suppose this last equation, which is of the form of 3·21, satisfied except by

$$x_{\mathbf{r}}' = x_{\mathbf{r}}'', \quad y_{\mathbf{r}}' = y_{\mathbf{r}}''$$

would be to suppose **a** and **b** collinear.

Let **f** be a vector in the vecplane built on **a** and **b** but not collinear with **b**. Then there is a relation

3·34 $$\mathbf{f} = x_{\mathbf{f}}\mathbf{a} + y_{\mathbf{f}}\mathbf{b},$$

and therefore any vector **r** which is expressible as $p\mathbf{f} + q\mathbf{b}$ is expressible as $(px_{\mathbf{f}})\,\mathbf{a} + (py_{\mathbf{f}} + q)\,\mathbf{b}$; that is, every vector that belongs to the vecplane built on **f** and **b** belongs to the vecplane built on **a** and **b**. But since **f** is not a multiple of **b**, the coefficient $x_{\mathbf{f}}$ in 3·34 is not zero, and 3·34 is equivalent to

3·35 $$\mathbf{a} = (1/x_{\mathbf{f}})\,\mathbf{f} + (-\,y_{\mathbf{f}}/x_{\mathbf{f}})\,\mathbf{b}\,;$$

hence **a** is a vector in the vecplane built on **f** and **b**, and therefore every vector that belongs to the vecplane built on **a** and **b** belongs to the vecplane built on **f** and **b**. Combining the two results, we conclude that the vecplane built on **f** and **b** coincides with the vecplane built on **a** and **b**. If further **g** is any vector in the vecplane built on **f** and **b** but not collinear with **f**, a repetition of the argument shews that the vecplane built on **f** and **g** coincides with the vecplane built on **f** and **b** and coincides therefore with the original vecplane. The proof fails if **f** is collinear with **b**, but in that case **f** is not collinear with **a**, and the vecplanes built on **a** and **b** and on **f** and **g** can both be compared with the vecplane built on **a** and **f**. Thus if **f** and **g** are any two vectors that belong to the vecplane built on **a** and **b** and are not themselves collinear, the vecplane built on **f** and **g** is identical with the vecplane built on **a** and **b**. In other words, the vecplane built on two vectors that are not collinear is the only vecplane that contains them both. The vecplane containing two vectors **a**, **b** is sometimes called simply the vecplane **ab**.

Obviously a vecplane that contains a proper vector **a** includes the whole of the vecline that contains **a**. The conclusion of the last paragraph can therefore be expressed in the form that if two veclines are distinct there is one and only one vecplane that includes them both. We must not be deceived by our vocabulary into supposing that two vecplanes necessarily have a common vecline, for this is not true. If the vecplane containing $\mathbf{a}'$ and $\mathbf{b}'$ and the vecplane containing $\mathbf{a}''$ and $\mathbf{b}''$ both contain a particular vector $\mathbf{r}$, there are alternative expressions $x'\mathbf{a}' + y'\mathbf{b}'$, $x''\mathbf{a}'' + y''\mathbf{b}''$ for this one vector, and equating them we have

$$x'\mathbf{a}' + y'\mathbf{b}' - x''\mathbf{a}'' - y''\mathbf{b}'' = \mathbf{0},$$

a linear relation between the vectors $\mathbf{a}'$, $\mathbf{b}'$, $\mathbf{a}''$, $\mathbf{b}''$, and since there are fewer than five of these vectors this relation implies either that the coefficients are all zero, that is, that $\mathbf{r}$ is the zero vector, or that the four vectors are not independent.

Ex. v. The vectors $(3, 0, 2, -1)$, $(-1, 2, 5, 3)$ are not collinear, and the vecplane built on them is the class of all tetrads of the form

$$(3x-y,\ 2y,\ 2x+5y,\ -x+3y).$$

To discover whether the particular vector $(-5, 4, 8, 7)$ is in this vecplane we must examine the set of equations

$$3x-y=-5,\ 2y=4,\ 2x+5y=8,\ -x+3y=7.$$

The first two equations in the set require $x=-1$, $y=2$, and with this pair of values of x and y the other two equations are satisfied. Hence the proposed vector does belong to the vecplane. In this example, *any* two of the four constituents *determine* the values of x and y; this example is *typical*, but the next example shews a peculiarity that sometimes presents itself.

Ex. vi. The typical member of the vecplane built on $(2, 1, -2, -1)$ and $(1, -1, -1, 3)$ is $(2x+y,\ x-y,\ -2x-y,\ -x+3y)$ and the set of equations which determines whether or not $(7, 2, -7, 4)$ has the requisite form is

$$2x+y=7,\ x-y=2,\ -2x-y=-7,\ -x+3y=4.$$

Here the first and third equations are virtually identical, but we have still three equations of which two determine values of x and y to be substituted in the third. To satisfy the first two equations in the set, we must have $x=3$, $y=1$, and with this pair of values the last equation is not satisfied; hence the vector suggested is not in the vecplane.

The number of distinct equations may reduce even to two, but to say that the four constituents gave only a single equation would be one way of expressing that the two vectors on which it had been proposed to build the vecplane were collinear and therefore unsuitable. For example, the *only* vector that combines with $(2, 1, -2, -1)$ to give $2x+y$ as a factor of every constituent is the collinear vector $(1, \frac{1}{2}, -1, -\frac{1}{2})$; the tetrads of the form

$$(2x+y,\ x+\tfrac{1}{2}y,\ -2x-y,\ -x-\tfrac{1}{2}y)$$

do not constitute a vecplane but compose only the vecline to which $(2, 1, -2, -1)$ and $(1, \frac{1}{2}, -1, -\frac{1}{2})$ both belong.

Ex. vii. The vectors $(-1, -2, 1, 4)$ and $(3, 0, -3, 2)$ belong to the vecplane built on $(2, 1, -2, -1)$ and $(1, -1, -1, 3)$, the first of them corresponding to $x=-1$, $y=1$, the second to $x=1$, $y=1$; also they are not collinear, for the only multiple of the first for which any constituent is zero is the zero vector. To verify that the vecplane built on $(-1, -2, 1, 4)$ and $(3, 0, -3, 2)$ is identical with that built on $(2, 1, -2, -1)$ and $(1, -1, -1, 3)$, we have to remark that any vector expressible as $x(2, 1, -2, -1)+y(1, -1, -1, 3)$ is expressible also as $(-\frac{1}{2}x+\frac{1}{2}y)(-1, -2, 1, 4)+(\frac{1}{2}x+\frac{1}{2}y)(3, 0, -3, 2)$ and that on the other hand any vector expressible as $X(-1, -2, 1, 4)+Y(3, 0, -3, 2)$ is expressible also as $(-X+Y)(2, 1, -2, -1)+(X+Y)(1, -1, -1, 3)$.

3·4. The relation between four vectors **a**, **b**, **c**, **d** when there are numbers a, b, c, d not all zero such that

3·41 $$a\mathbf{a}+b\mathbf{b}+c\mathbf{c}+d\mathbf{d}=\mathbf{0}$$

is expressed by describing the vectors as *cospatial*. If **a**, **b**, **c** are coplanar, we can take d as zero and reduce the equation 3·41 to the form of 3·31 which can then by hypothesis be satisfied: if three vectors are coplanar, every vector is cospatial with them. If **a**, **b**, **c** are not coplanar, it is impossible for d to be zero if 3·41 is satisfied, and therefore 3·41 can be replaced by

3·42 $$\mathbf{d}=(-a/d)\,\mathbf{a}+(-b/d)\,\mathbf{b}+(-c/d)\,\mathbf{c}.$$

The class of vectors cospatial with three vectors **a**, **b**, **c** that are not coplanar is the *vecspace* built on **a**, **b**, **c**, or briefly the vecspace **abc**. Corresponding to every vector **r** in this vecspace there is a set of coefficients $(x_{\mathbf{r}}, y_{\mathbf{r}}, z_{\mathbf{r}})$ such that

3·43 $$\mathbf{r}=x_{\mathbf{r}}\mathbf{a}+y_{\mathbf{r}}\mathbf{b}+z_{\mathbf{r}}\mathbf{c},$$

and there is only one set associated with each vector in the vecspace, for the existence of two sets $(x_{\mathbf{r}}', y_{\mathbf{r}}', z_{\mathbf{r}}')$ and $(x_{\mathbf{r}}'', y_{\mathbf{r}}'', z_{\mathbf{r}}'')$ for the same vector **r** implies the equation

$$(x_{\mathbf{r}}'-x_{\mathbf{r}}'')\,\mathbf{a}+(y_{\mathbf{r}}'-y_{\mathbf{r}}'')\,\mathbf{b}+(z_{\mathbf{r}}'-z_{\mathbf{r}}'')\,\mathbf{c}=\mathbf{0},$$

and since this equation is of the form of 3·31 and **a**, **b**, **c** are by hypothesis not coplanar, the coefficients $x_{\mathbf{r}}'-x_{\mathbf{r}}''$, $y_{\mathbf{r}}'-y_{\mathbf{r}}''$, $z_{\mathbf{r}}'-z_{\mathbf{r}}''$ must vanish individually.

If **f** is a vector in the vecspace built on **a**, **b**, **c** but not coplanar with **b** and **c**, the relation

3·44 $$\mathbf{f}=x_{\mathbf{f}}\mathbf{a}+y_{\mathbf{f}}\mathbf{b}+z_{\mathbf{f}}\mathbf{c}$$

is equivalent to

3·45 $$\mathbf{a}=(1/x_{\mathbf{f}})\,\mathbf{f}+(-y_{\mathbf{f}}/x_{\mathbf{f}})\,\mathbf{b}+(-z_{\mathbf{f}}/x_{\mathbf{f}})\,\mathbf{c},$$

and it follows by reasoning parallel to that in section 3·3 that the vecspace built on **a**, **b**, **c** coincides with that built on **f**, **b**, **c**. If then **g** is in this vecspace and is not coplanar with **f** and **c**, a second application of the argument proves the same vecspace to be that built on **f**, **g**, **c**; and finally if **h** is in the vecspace and is not coplanar with **f** and **g**, this vecspace is identified with that built on **f**, **g**, **h**. If **g** is a proper vector coplanar with **f** and **c**, it can not be coplanar also with **f** and **b**, if **f**, **b**, **c** are not coplanar; hence in this case we may pass from the vecspace built on **f**, **b**, **c** to that built on **f**, **g**, **b** and thence to that built on **f**, **g**, **h**. If **f** is coplanar with **b** and **c** but not with **a** and **c**, we may pass first from the vecspace built on **a**, **b**, **c** to that built on **a**, **f**, **c**. And lastly if **f** is coplanar both with **b** and **c** and with **a** and **c**, then **f** belongs to the vecline containing **c** and is *not* coplanar with **a** and **b**. Thus we can draw a universal conclusion: if **f**, **g**, **h** are any three vectors that are in the vecspace built on **a**, **b**, **c** but are not themselves coplanar, then the vecspace built on **f**, **g**, **h** is the same as the vecspace built on **a**, **b**, **c**. That is, the vecspace built on three vectors that are not coplanar is the only vecspace that contains them all.

A vecspace that contains a proper vector **a** includes the whole of the vecline to which **a** belongs, and a vecspace that includes two distinct veclines includes the vecplane of which these veclines form part. If three veclines are not coplanar there is one and only one vecspace that includes them all. If a given vecline is not included in a given vecplane, any two distinct veclines in the vecplane form with this vecline a set of three veclines that are not coplanar, and the vecspace that includes these three veclines includes both the given vecline and the given vecplane and is the only vecspace that does include them both: if a vecline is not part of a vecplane there is one and only one vecspace that includes both the vecline and the vecplane.

Ex. viii. The vectors $(2, 1, -2, -1)$, $(1, -1, -1, 3)$, $(7, 2, -7, 4)$, as has been seen in Example vi (p. 11), are not coplanar; the vecspace built on them is the class of tetrads of the form

$$(2x+y+7z,\ x-y+2z,\ -2x-y-7z,\ -x+3y+4z).$$

Among the members of this vecspace are $(10, 2, -10, 6)$, $(-9, -6, 9, 0)$ and $(-2, -4, 2, 4)$ which are not coplanar, and by solving the set of equations

$$2x+y+7z=10X-9Y-2Z, \qquad x-y+2z=2X-6Y-4Z,$$
$$-2x-y-7z=-10X+9Y+2Z, \qquad -x+3y+4z=6X+4Z$$

both in the form

$$X=\tfrac{3}{2}x-\tfrac{1}{2}y,\quad Y=2x-y-z,\quad Z=-\tfrac{5}{2}x+\tfrac{3}{2}y+z$$

and in the form

$$x=X+Y+Z,\quad y=X+3Y+3Z,\quad z=X-2Y-Z$$

we shew in the clearest light the identity of the original vecspace with that built on the three vectors last mentioned.

To say that a vecspace and a vecline have in general no common members except the zero vector is merely to affirm that one vecspace does not contain all the vectors there are, for the equation

$$x_{\mathbf{r}}'\mathbf{a}' = x_{\mathbf{r}}''\mathbf{a}'' + y_{\mathbf{r}}''\mathbf{b}'' + z_{\mathbf{r}}''\mathbf{c}''$$

which expresses that the vector **r** belongs both to the vecline built on $\mathbf{a}'$ and to the vecspace built on $\mathbf{a}''$, $\mathbf{b}''$, $\mathbf{c}''$ expresses also, unless $x_{\mathbf{r}}'$ is zero, that $\mathbf{a}'$ belongs to the vecspace.

Suppose that $\mathbf{a}'$, $\mathbf{b}'$ are two vectors that are not collinear and that $\mathbf{a}''$, $\mathbf{b}''$, $\mathbf{c}''$ are three vectors that are not coplanar. There is an effective linear relation between the five vectors, and this can be exhibited in the form

$$x'\mathbf{a}' + y'\mathbf{b}' = x''\mathbf{a}'' + y''\mathbf{b}'' + z''\mathbf{c}'',$$

in which form it shews that the vecplane $\mathbf{a}'\mathbf{b}'$ and the vecspace $\mathbf{a}''\mathbf{b}''\mathbf{c}''$ have in common the vector **r** for which $x'\mathbf{a}' + y'\mathbf{b}'$ and $x''\mathbf{a}''+y''\mathbf{b}''+z''\mathbf{c}''$ are alternative expressions. This vector **r** can not be merely the zero vector, for by hypothesis neither of the conditions

$$x'\mathbf{a}' + y'\mathbf{b}' = \mathbf{0},\quad x''\mathbf{a}'' + y''\mathbf{b}'' + z''\mathbf{c}'' = \mathbf{0}$$

can be satisfied. Thus a vecplane and a vecspace necessarily have one vecline in common. In general, a vecplane and a vecspace have only one common vecline, for if the vecspace includes two distinct veclines belonging to the vecplane, it includes the whole vecplane.

Ex. ix. The vecspace of Example viii and the vecplane built on (3, −2, 1, 0) and (1, 1, −5, 9) have in common the vector (4, −1, −4, 9), for this vector is expressible both as −2 (2, 1, −2, −1) + (1, −1, −1, 3) + (7, 2, −7, 4) and as (3, −2, 1, 0) + (1, 1, −5, 9); they have in common therefore every multiple of (4, −1, −4, 9), but they have no other common members, for the vecplane is not wholly included in the vecspace.

Given a vecspace **fgh** and a vecspace built on three vectors **a**, **b**, **c**, there is certainly one vecline L common to **fgh** and the vecplane **bc**, there is certainly one vecline M common to **fgh** and the vecplane **ca**, and there is certainly one vecline N common to **fgh** and the vecplane **ab**. Since the only vecline common to the vecplane **ca** and the vecplane **ab** is the vecline containing **a**, and since **a** is not in the vecplane **bc**, the vecplanes **bc**, **ca**, **ab** have no common vecline, and therefore of the three veclines L, M, N two are certainly distinct. Hence the two vecspaces **fgh**, **abc** have in common two distinct veclines, and therefore have in common the whole of the vecplane that includes these veclines; if the two vecspaces had in common any vecline outside this vecplane they would coincide. Thus if two vecspaces are distinct, the vectors common to them compose a definite vecplane.

Ex. x. The vecspace of Example viii above and the vecspace built on $(-5, -9, 9, -1)$, $(3, -2, 1, 0)$, $(1, 1, -5, 9)$ have in common the coplanar vectors $(4, -1, -4, 9)$, $(-1, -2, 1, 2)$, $(8, 7, -8, 1)$, no two of which are collinear. Since the two vecspaces are not identical, the only vectors common to them are those composing the vecplane that contains the three vectors mentioned.

The vecplane common to two given vecspaces may be included in a third given vecspace, but in general has in common with this third vecspace only the vectors composing a single vecline. Hence in general the vectors common to three given vecspaces compose a definite vecline, but it is possible for three vecspaces to include a common vecplane without any two of the vecspaces coinciding.

If two vecplanes are defined independently, there is not as a rule any one vecspace that includes them both. In fact, the vecplane containing **a** and **b** and the vecplane containing **c** and **d** are in one vecspace if and only if the four vectors **a**, **b**, **c**, **d** are cospatial, and this condition is a restriction on the vectors. The proposition at the end of section 3·3 may be expressed in the form that two vecplanes have a vecline in common if and only if there is a vecspace in which they are both included.

Ex. xi. Between the four vectors

$$(2, 1, -2, -1),\ (1, -1, -1, 3),\ (7, 2, -7, 4),\ (3, 2, -3, 0),$$

which we will denote by **k**, **l**, **m**, **n**, there is the linear relation

$$\mathbf{k}+3\mathbf{l}-2\mathbf{m}+3\mathbf{n}=\mathbf{0},$$

from which we deduce not only that the four vectors are cospatial, but also that the vecplanes **kl** and **mn** have in common a vector expressible both as $\mathbf{k}+3\mathbf{l}$ and as $2\mathbf{m}-3\mathbf{n}$, which is $(5, -2, -5, 8)$, that the vecplanes **km** and **ln** have in common a vector expressible both as $-\frac{1}{3}\mathbf{k}+\frac{2}{3}\mathbf{m}$ and as $\mathbf{l}+\mathbf{n}$, which is (4, 1, −4, 3), and that the vecplanes **kn** and **lm** have in common a vector expressible both as **k**+3**n** and as −3**l**+2**m**, which is (11, 7, −11, −1).

3·5. We have said that between any five vectors **a**, **b**, **c**, **d**, **e** there is at least one effective relation of the form

3·51 $$a\mathbf{a} + b\mathbf{b} + c\mathbf{c} + d\mathbf{d} + e\mathbf{e} = \mathbf{0}.$$

If in this relation e is zero, the equation reduces to the form of 3·41 and asserts that the four vectors **a**, **b**, **c**, **d** are cospatial. If these vectors are not cospatial, e is not zero and the relation is equivalent to

3·52 $$\mathbf{e} = (-a/e)\,\mathbf{a} + (-b/e)\,\mathbf{b} + (-c/e)\,\mathbf{c} + (-d/e)\,\mathbf{d}.$$

Thus if **a**, **b**, **c**, **d** are any four vectors that are not cospatial, to any vector **r** there corresponds a set of coefficients $(x_\mathbf{r}, y_\mathbf{r}, z_\mathbf{r}, t_\mathbf{r})$ such that

3·53 $$\mathbf{r} = x_\mathbf{r}\mathbf{a} + y_\mathbf{r}\mathbf{b} + z_\mathbf{r}\mathbf{c} + t_\mathbf{r}\mathbf{d}.$$

The set is proved to be unique by an argument precisely similar to one used in 3·3 and 3·4. The coefficients in 3·53 are called the coefficients or components of **r** in the *vecframe* **abcd**. The absolute components of **r** are given in terms of the coefficients in the vecframe by such formulae as

3·54 $$\xi_\mathbf{r} = x_\mathbf{r}\,\xi_\mathbf{a} + y_\mathbf{r}\,\xi_\mathbf{b} + z_\mathbf{r}\,\xi_\mathbf{c} + t_\mathbf{r}\,\xi_\mathbf{d};$$

to find the coefficients in terms of the absolute components it is necessary to treat the four formulae of the type of 3·54 as simultaneous equations with the four coefficients as variables.

Ex. xii. In the vecframe formed of the vectors **l**, **m**, **n** of Example xi above together with the vector (1, 2, 3, 4), which is not cospatial with them, the coefficients of (3, 3, 1, 3) are −3, 2, −3, 1, for identically

(3, 3, 1, 3) = −3 (1, −1, −1, 3) + 2 (7, 2, −7, 4) − 3 (3, 2, −3, 0) + (1, 2, 3, 4).

4. THE DETERMINATION OF VECSPACES BY EQUATIONS

4·1. Two vectors **a**, **b** are said to be *perpendicular* to each other if

4·11 $$\xi_\mathbf{a}\xi_\mathbf{b} + \eta_\mathbf{a}\eta_\mathbf{b} + \zeta_\mathbf{a}\zeta_\mathbf{b} + \tau_\mathbf{a}\tau_\mathbf{b} = 0.$$

The zero vector is perpendicular to every vector, itself included. If two proper vectors **a**, **b** are perpendicular, every multiple of **a** is perpendicular to every multiple of **b**, and therefore the vecline containing **a** and the vecline containing **b** may be described as perpendicular veclines. We can assert that two vectors are perpendicular if and only if they belong to perpendicular veclines, for if **b** is proper, the zero vector belongs to veclines perpendicular to the vecline containing **b** as well as to veclines that are not perpendicular to this vecline.

No vector other than the zero vector is self-perpendicular*, and no vecline is self-perpendicular.

A vecline that is perpendicular to each of two distinct veclines is perpendicular to every vecline in the vecplane containing these two veclines, and may be described as perpendicular to the vecplane itself.

Ex. xiii. Any two of the four vectors

$$(10, -3, -4, 10),\ (-5, 6, 8, 10),\ (0, 4, -3, 0),\ (10, 6, 8, -5)$$

are perpendicular to each other; so also are any two of the four vectors

$$(2, -1, 0, 2),\ (-1, 2, 4, 2),\ (-4, 8, -9, 8),\ (2, 2, 0, -1).$$

4·2. The proposition that there is at least one effective linear relation between any five vectors, which we have used in 3·4 and in 3·5, is a particular case of the general proposition that *there is at least one effective linear relation between any $n+1$ sets of n numbers*, that is, that whatever the values of the constituent numbers in the sets

$$(X_1^{(1)}, X_2^{(1)}, \ldots, X_n^{(1)}),\ (X_1^{(2)}, X_2^{(2)}, \ldots, X_n^{(2)}),\ \ldots,\ (X_1^{(n+1)}, X_2^{(n+1)}, \ldots, X_n^{(n+1)}),$$

there is at least one set of numbers $(\lambda_{(1)}, \lambda_{(2)}, \ldots, \lambda_{(n+1)})$ not all zero such that

4·21
$$\begin{cases} \lambda_{(1)} X_1^{(1)} + \lambda_{(2)} X_1^{(2)} + \ldots + \lambda_{(n+1)} X_1^{(n+1)} = 0, \\ \lambda_{(1)} X_2^{(1)} + \lambda_{(2)} X_2^{(2)} + \ldots + \lambda_{(n+1)} X_2^{(n+1)} = 0, \\ \ldots\ldots\ldots\ldots\ldots\ldots\ldots\ldots\ldots\ldots\ldots\ldots, \\ \lambda_{(1)} X_n^{(1)} + \lambda_{(2)} X_n^{(2)} + \ldots + \lambda_{(n+1)} X_n^{(n+1)} = 0. \end{cases}$$

Suppose that **r** is a vector in the vecspace **abc**. Then there are numbers $x_{\mathbf{r}}, y_{\mathbf{r}}, z_{\mathbf{r}}$ such that

$$\mathbf{r} = x_{\mathbf{r}}\mathbf{a} + y_{\mathbf{r}}\mathbf{b} + z_{\mathbf{r}}\mathbf{c},$$

that is, such that

4·22
$$\begin{cases} \xi_{\mathbf{r}} = x_{\mathbf{r}}\xi_{\mathbf{a}} + y_{\mathbf{r}}\xi_{\mathbf{b}} + z_{\mathbf{r}}\xi_{\mathbf{c}}, \\ \eta_{\mathbf{r}} = x_{\mathbf{r}}\eta_{\mathbf{a}} + y_{\mathbf{r}}\eta_{\mathbf{b}} + z_{\mathbf{r}}\eta_{\mathbf{c}}, \\ \zeta_{\mathbf{r}} = x_{\mathbf{r}}\zeta_{\mathbf{a}} + y_{\mathbf{r}}\zeta_{\mathbf{b}} + z_{\mathbf{r}}\zeta_{\mathbf{c}}, \\ \tau_{\mathbf{r}} = x_{\mathbf{r}}\tau_{\mathbf{a}} + y_{\mathbf{r}}\tau_{\mathbf{b}} + z_{\mathbf{r}}\tau_{\mathbf{c}}. \end{cases}$$

* If space is constructed of sets of *complex* numbers this statement is not true, and self-perpendicular proper vectors are of the greatest importance. I have not scrupled to use the principle that no proper vector is self-perpendicular to abbreviate proofs even of some propositions that are true of complex space, but alternative proofs of these propositions are to be found in the appendix.

Looking on $(\xi_{\mathbf{a}}, \xi_{\mathbf{b}}, \xi_{\mathbf{c}})$, $(\eta_{\mathbf{a}}, \eta_{\mathbf{b}}, \eta_{\mathbf{c}})$, $(\zeta_{\mathbf{a}}, \zeta_{\mathbf{b}}, \zeta_{\mathbf{c}})$, $(\tau_{\mathbf{a}}, \tau_{\mathbf{b}}, \tau_{\mathbf{c}})$ as four triads, we infer from the algebraic theorem just enunciated that there is a set of numbers $(\lambda, \mu, \nu, \varpi)$ not all zero such that

4·23
$$\begin{cases} \lambda\xi_{\mathbf{a}} + \mu\eta_{\mathbf{a}} + \nu\zeta_{\mathbf{a}} + \varpi\tau_{\mathbf{a}} = 0, \\ \lambda\xi_{\mathbf{b}} + \mu\eta_{\mathbf{b}} + \nu\zeta_{\mathbf{b}} + \varpi\tau_{\mathbf{b}} = 0, \\ \lambda\xi_{\mathbf{c}} + \mu\eta_{\mathbf{c}} + \nu\zeta_{\mathbf{c}} + \varpi\tau_{\mathbf{c}} = 0; \end{cases}$$

taking a definite set of numbers satisfying this set of equations, we have immediately from 4·22 and 4·23

4·24
$$\lambda\xi_{\mathbf{r}} + \mu\eta_{\mathbf{r}} + \nu\zeta_{\mathbf{r}} + \varpi\tau_{\mathbf{r}} = 0,$$

whatever the values of $x_{\mathbf{r}}, y_{\mathbf{r}}, z_{\mathbf{r}}$. Since there is no reference to **r** in the set of equations 4·23, the values of $\lambda, \mu, \nu, \varpi$ are constants dependent on the vecspace but not on **r**. If the relation 4·24 was satisfied by the absolute components of a vector **d** not cospatial with **a**, **b**, **c** it would be satisfied by those of any vector **r** expressible in the form $x_{\mathbf{r}}\mathbf{a} + y_{\mathbf{r}}\mathbf{b} + z_{\mathbf{r}}\mathbf{c} + t_{\mathbf{r}}\mathbf{d}$, that is, by those of any vector whatever; in other words, $\lambda, \mu, \nu, \varpi$ would be four numbers such that $\lambda a + \mu b + \nu c + \varpi d$ was zero for every possible set of values of a, b, c, d, and it is obvious that no four numbers that are not all zero can have this property—in fact, we can avoid the conclusion

$$\lambda a + \mu b + \nu c + \varpi d = 0$$

simply* by taking a, b, c, d equal respectively to $\lambda, \mu, \nu, \varpi$. Hence 4·24 is a relation satisfied by the absolute components of **r** *if and only if* **r** belongs to the vecspace **abc**.

4·3. The coefficients $\lambda, \mu, \nu, \varpi$ appearing in 4·24 can be regarded as the absolute components of some definite proper vector, and we see that associated with every vecspace there is at least one vecline such that every vector in the vecspace is perpendicular to every vector in the vecline. If there were two distinct veclines with this relation to the vecspace, then every vecline in the vecplane containing them would be perpendicular to every vecline in the vecspace, and therefore the vecline common to the vecspace and this vecplane would be self-perpendicular. It follows† that there is one and only one vecline that is perpendicular to a given vecspace, and it follows further that there is no equation of the form of 4·24 which is satisfied by the absolute components of *every* vector

* See Appendix, p. 53. † See Appendix, p. 53.

in the vecspace **abc** that is not merely a multiple of the actual equation in 4·24. If **n** is any proper vector in the vecline perpendicular to a given vecspace, we may call the equation

4·31 $$\xi_{\mathbf{n}}\xi_{\mathbf{r}}+\eta_{\mathbf{n}}\eta_{\mathbf{r}}+\zeta_{\mathbf{n}}\zeta_{\mathbf{r}}+\tau_{\mathbf{n}}\tau_{\mathbf{r}}=0$$

the equation of the vecspace, since the vector **r** belongs to the vecspace if and only if this equation is satisfied.

If two vecspaces are distinct, the veclines perpendicular to them are distinct also. A vecspace and the vecline perpendicular to it have no common member other than the zero vector.

Ex. xiv. The vecspace of Example viii (p. 13) obviously has the equation

$$\xi+\zeta=0.$$

If ξ, η, τ, have any values whatever, the set of equations

$$2x+y+7z=\xi,\quad x-y+2z=\eta,\quad -x+3y+4z=\tau$$

is satisfied by

$$x=\tfrac{5}{6}\xi-\tfrac{17}{12}\eta-\tfrac{3}{4}\tau,\quad y=\tfrac{1}{2}\xi-\tfrac{5}{4}\eta-\tfrac{1}{4}\tau,\quad z=-\tfrac{1}{6}\xi+\tfrac{7}{12}\eta+\tfrac{1}{4}\tau,$$

and no condition except $\xi+\zeta=0$ is necessary to secure that with these values of x, y, z

$$-2x-y-7z=\zeta.$$

That is to say, the vector (ξ, η, ζ, τ) is expressible as

$$x(2, 1, -2, -1)+y(1, -1, -1, 3)+z(7, 2, -7, 4)$$

if and only if the equation

$$\xi+\zeta=0$$

is satisfied.

Ex. xv. To find the equation of the vecspace introduced in Example x (p. 15) is to *eliminate* x, y, z between the four equations in the set

$$-5x+3y+z=\xi,\quad -9x-2y+z=\eta,\quad 9x+y-5z=\zeta,\quad -x+9z=\tau,$$

that is, to calculate the values of x, y, z that satisfy three of the equations and to discover the result of substituting these values in the fourth equation. From the last three equations we have

$$8x=\eta+2\zeta+\tau,\quad 18y=-19\eta-20\zeta-9\tau,\quad 72z=\eta+2\zeta+9\tau,$$

and the required equation is easily found to be

$$9\xi+34\eta+41\zeta+18\tau=0.$$

Any vecplane p can be regarded as the common part of two vecspaces that are not identical. If **m** is a proper vector perpendicular to one of these vecspaces and **n** a proper vector perpendicular to the other, every vector in the vecplane is perpendicular to both **m** and **n** and is therefore perpendicular to every vector in the vecplane **mn**. Moreover, no vector that does not belong to **mn** can be perpendicular to every vector in p, for if $\mathbf{l}$ is any vector not in $\mathbf{mn}$, the vectors perpendicular to the three vectors $\mathbf{l}$, $\mathbf{m}$, $\mathbf{n}$ occupy a single vecline. Thus the vectors perpendicular to one vecplane compose a second vecplane.

For the vector $\mathbf{r}$ to belong to the vecplane common to the vecspaces to which $\mathbf{m}$ and $\mathbf{n}$ are perpendicular, the absolute components of $\mathbf{r}$ must satisfy the pair of equations

4·32
$$\begin{cases} \xi_{\mathbf{m}}\xi_{\mathbf{r}} + \eta_{\mathbf{m}}\eta_{\mathbf{r}} + \zeta_{\mathbf{m}}\zeta_{\mathbf{r}} + \tau_{\mathbf{m}}\tau_{\mathbf{r}} = 0, \\ \xi_{\mathbf{n}}\xi_{\mathbf{r}} + \eta_{\mathbf{n}}\eta_{\mathbf{r}} + \zeta_{\mathbf{n}}\zeta_{\mathbf{r}} + \tau_{\mathbf{n}}\tau_{\mathbf{r}} = 0, \end{cases}$$

and this is therefore *a* pair of equations of the vecplane, but there is nothing to distinguish this pair of equations from the pair associated with any other pair of vectors in the vecplane $\mathbf{mn}$, and there is no one pair of equations that can be regarded as *the* pair of equations of a particular vecplane.

It is always possible to find for a given vecplane equations in which the constituents ξ, η, ζ, τ do not all occur, and these may be regarded as standard equations. In general there are four distinct standard equations, of which any two may be deduced from the others, but in special cases one or two of the equations may fail altogether or as many as three of them may coincide; for this reason we can not specify in advance a particular pair of standard equations to be used.

Ex. xvi. The vecplane common to the vecspaces of the last two examples has the pair of equations

$$\xi+\zeta=0, \quad 9\xi+34\eta+41\zeta+18\tau=0.$$

A set of standard equations is therefore

$$17\eta+16\zeta+9\tau=0, \quad \xi+\zeta=0, \quad -16\xi+17\eta+9\tau=0, \quad \xi+\zeta=0,$$

of which the second and fourth are identical.

If the equation of a vecspace does not involve ξ, the vector $(\xi, 0, 0, 0)$ belongs to the vecspace whatever the value of ξ. Thus the first standard equation of a vecplane is the equation of a vecspace that includes both the given vecplane and the vecline built on $(1, 0, 0, 0)$, and the equation disappears only if the vecline is part of the vecplane. The second and the fourth of the standard equations in the present example coincide because the vecplane built on $(1, 0, 0, 0)$ and $(0, 0, 1, 0)$ is cospatial with the vecplane under consideration.

To find a standard equation for a vecplane directly from a pair of vectors on which the vecplane is built, is to find an equation of a vecspace built on three given vectors, namely, the two that determine the vecplane together with one of the four fundamental vectors

$$(1, 0, 0, 0), \quad (0, 1, 0, 0), \quad (0, 0, 1, 0), \quad (0, 0, 0, 1).$$

Thus the first standard equation of the vecplane built on $(4, -1, -4, 9)$ and $(-1, -2, 1, 2)$ is to be found by eliminating x, y, z from the set of equations

$$4x-y+z=\xi, \quad -x-2y=\eta, \quad -4x+y=\zeta, \quad 9x+2y=\tau.$$

If x and y satisfy the last three of these equations, z can be adjusted from the first of them, whatever the value of ξ; thus ξ is not involved in the result, which is easily seen to be $17\eta+16\zeta+9\tau=0$. The second standard equation of the same vecplane is the eliminant of

$$4x-y=\xi, \quad -x-2y+z=\eta, \quad -4x+y=\zeta, \quad 9x+2y=\tau,$$

and is $\xi+\zeta=0$. Thus the present example is in agreement with Example x (p. 15).

By considering a vecline as the common part of three vecspaces that do not include a common vecplane we obtain the correlatives of the results already found in this section. We see that vectors that are perpendicular to a given vecline compose a definite vecspace. From this it follows that not only has every vecspace an equation of the form of 4·31 but every equation of this form corresponds to some definite vecspace, and that not only does every vecplane give rise to pairs of equations such as 4·32 but every such pair of equations is associated with a definite vecplane if the two equations are really distinct, that is, if the vectors **m**, **n** are not collinear. The absolute components of a vector **r** in a vecline satisfy a set of three equations such as

4·33
$$\begin{cases} \xi_{\mathbf{l}}\xi_{\mathbf{r}} + \eta_{\mathbf{l}}\eta_{\mathbf{r}} + \zeta_{\mathbf{l}}\zeta_{\mathbf{r}} + \tau_{\mathbf{l}}\tau_{\mathbf{r}} = 0, \\ \xi_{\mathbf{m}}\xi_{\mathbf{r}} + \eta_{\mathbf{m}}\eta_{\mathbf{r}} + \zeta_{\mathbf{m}}\zeta_{\mathbf{r}} + \tau_{\mathbf{m}}\tau_{\mathbf{r}} = 0, \\ \xi_{\mathbf{n}}\xi_{\mathbf{r}} + \eta_{\mathbf{n}}\eta_{\mathbf{r}} + \zeta_{\mathbf{n}}\zeta_{\mathbf{r}} + \tau_{\mathbf{n}}\tau_{\mathbf{r}} = 0, \end{cases}$$

but this set can be modified without any change in the vecline represented by it. A standard equation of a vecline is an equation in which not more than two of the absolute components are involved. There can not be fewer than three distinct standard equations, and as a rule there are six of which any three determine the vecline.

Ex. xvii. The standard equations of the vecline built on $(3, 1, -5, -2)$ are

$$\xi-3\eta=0, \quad 5\xi+3\zeta=0, \quad 2\xi+3\tau=0, \quad 5\eta+\zeta=0, \quad 2\eta+\tau=0, \quad 2\zeta-5\tau=0.$$

Each of these defines a vecspace containing the given vector and two of the four fundamental vectors. The equation $2\xi+\eta+\zeta+\tau=0$ can be combined with other equations to specify the vecline, but it is not a *standard* equation.

5. MEASUREMENT; DIRECTIONS AND ANGLES

5·1. A number r which is such that

5·11 $$r^2 = \xi_{\mathbf{r}}^2 + \eta_{\mathbf{r}}^2 + \zeta_{\mathbf{r}}^2 + \tau_{\mathbf{r}}^2$$

is called a *measure* of the vector $\mathbf{r}$. The zero vector has zero for its only measure. A proper vector has two* measures, one positive and the other negative. A vector with one of its measures associated with it is a *measured vector.*

A *direction* is† an ordered set of four numbers the sum of whose squares is unity. That is, (a, b, c, d) is a direction if

5·12 $$a^2 + b^2 + c^2 + d^2 = 1.$$

If (a, b, c, d) is a direction, so also is $(-a, -b, -c, -d)$, which is called the direction opposite to (a, b, c, d).

Ex. xviii. The measures of $(-7, 2, 4, -10)$ are 13 and -13, and the measures of $(f^2, fg, -g^2, fg)$ are f^2+g^2 and $-(f^2+g^2)$.

If r is a measure of a proper vector $\mathbf{r}$, then $\mathbf{r}/r$ is a direction, and this direction is naturally called the direction of the measured vector. If the same vector $\mathbf{r}$ is measured by $-r$, the direction is reversed, becoming $-\mathbf{r}/r$. A proper vector has two directions, and is said to be measured *in* one or other of these directions.

Ex. xix. The vectors $(-7, 2, 4, -10)$, $(1, 2, -4, 2)$ have the measures -13, -5 in the directions $(\frac{7}{13}, -\frac{2}{13}, -\frac{4}{13}, \frac{10}{13})$, $(-\frac{1}{5}, -\frac{2}{5}, \frac{4}{5}, -\frac{2}{5})$.

Since a vector is the product of either of its directions by its measure in that direction, two proper vectors can have the same pair of directions only if they are collinear. On the other hand, if $\mathbf{r}$ and $\mathbf{s}$ are proper collinear vectors, there is a number k such that

5·13 $$\mathbf{s} = k\mathbf{r},$$

and since the four equations

5·14 $$\xi_{\mathbf{s}} = k\xi_{\mathbf{r}}, \quad \eta_{\mathbf{s}} = k\eta_{\mathbf{r}}, \quad \zeta_{\mathbf{s}} = k\zeta_{\mathbf{r}}, \quad \tau_{\mathbf{s}} = k\tau_{\mathbf{r}},$$

involved in 5·13 imply

5·15 $$\xi_{\mathbf{s}}^2 + \eta_{\mathbf{s}}^2 + \zeta_{\mathbf{s}}^2 + \tau_{\mathbf{s}}^2 = k^2 r^2$$

if r satisfies 5·11, the measures of $\mathbf{s}$ are kr and $-kr$, and one of

* See Appendix, p. 52.

† In an alternative process, the meaning of the assertion that two measured vectors are codirectional is first defined, in an obvious manner, and a direction is then defined as a class of codirectional measured vectors. Logically this course is far superior to that in the text, but it involves a fresh difficulty for anyone unaccustomed to Frege-Russell definitions, and with reluctance I have avoided it. If coordinates are complex numbers, the method in the text can not be adopted.

the directions of $\mathbf{s}$ is $\mathbf{s}/kr$ which is the direction $\mathbf{r}/r$. Thus collinear proper vectors possess the same pair of directions, and these directions may be regarded as belonging to the vecline that contains the vectors. It is consistent to regard the zero vector $\mathbf{0}$ as having all directions, since it belongs to every vecline; the measure of $\mathbf{0}$ in any direction is zero.

A vector whose measure is unity is sometimes called a unit vector or a *radial*. A radial is to be considered as having only one direction, and in fact we have drawn no distinction in definition between a radial and its direction. But a unit vector, like any other vector, is often expressed as the sum of a number of parts, and we avoid an unfamiliar turn of speech if we refrain from speaking of a *direction* as decomposed into a sum of vectors.

5·2. Whatever the values of the eight numbers involved, we have identically

5·21 $$(\xi\mathbf{u}^2+\eta\mathbf{u}^2+\zeta\mathbf{u}^2+\tau\mathbf{u}^2)(\xi\mathbf{v}^2+\eta\mathbf{v}^2+\zeta\mathbf{v}^2+\tau\mathbf{v}^2) - (\xi\mathbf{u}\,\xi\mathbf{v}+\eta\mathbf{u}\,\eta\mathbf{v}+\zeta\mathbf{u}\,\zeta\mathbf{v}+\tau\mathbf{u}\,\tau\mathbf{v})^2$$
$$=(\xi\mathbf{u}\,\eta\mathbf{v}-\eta\mathbf{u}\,\xi\mathbf{v})^2+(\xi\mathbf{u}\,\zeta\mathbf{v}-\zeta\mathbf{u}\,\xi\mathbf{v})^2+(\xi\mathbf{u}\,\tau\mathbf{v}-\tau\mathbf{u}\,\xi\mathbf{v})^2 +(\eta\mathbf{u}\,\zeta\mathbf{v}-\zeta\mathbf{u}\,\eta\mathbf{v})^2+(\eta\mathbf{u}\,\tau\mathbf{v}-\tau\mathbf{u}\,\eta\mathbf{v})^2+(\zeta\mathbf{u}\,\tau\mathbf{v}-\tau\mathbf{u}\,\zeta\mathbf{v})^2.$$

The expression on the right, being the sum of a number of squares, can not be negative. Hence if $\mathbf{u}$ and $\mathbf{v}$ are *directions*, so that

$$(\xi\mathbf{u}^2+\eta\mathbf{u}^2+\zeta\mathbf{u}^2+\tau\mathbf{u}^2)(\xi\mathbf{v}^2+\eta\mathbf{v}^2+\zeta\mathbf{v}^2+\tau\mathbf{v}^2)$$

is unity,

$$\xi\mathbf{u}\,\xi\mathbf{v}+\eta\mathbf{u}\,\eta\mathbf{v}+\zeta\mathbf{u}\,\zeta\mathbf{v}+\tau\mathbf{u}\,\tau\mathbf{v}$$

is not less than -1 and not greater than $+1$, and there are angles which have

$$\xi\mathbf{u}\,\xi\mathbf{v}+\eta\mathbf{u}\,\eta\mathbf{v}+\zeta\mathbf{u}\,\zeta\mathbf{v}+\tau\mathbf{u}\,\tau\mathbf{v}$$

for their cosine. These angles are called angles *between* **u** *and* **v**, or between any measured vectors with the directions **u** and **v**. If δ is any one angle between **u** and **v**, every angle between **u** and **v** has one of the two forms $2n\pi+\delta$, $2n\pi-\delta$, where n is integral; angles of the one form have the sine $\sin\delta$, and angles of the other the sine $-\sin\delta$.

It will be noticed that two directions are perpendicular, in the sense of the word introduced in 4·1, if there is a right angle between the directions.

Ex. xx. The angles a direction makes with itself are the even multiples of π, and the angles between a direction and its reverse are the odd multiples of π.

5·3. In general we can only accept *all* the angles with the appropriate cosine as angles between two directions, but if four directions are coplanar we can establish a definite relation between one of the possible sines for angles between two of the directions and one of the possible sines for angles between the other two. Let $\mathbf{a}$, $\mathbf{b}$ be unit vectors that are not collinear, and let $\mathbf{u}$, $\mathbf{v}$ be unit vectors coplanar with $\mathbf{a}$ and $\mathbf{b}$, given by

5·31 $$\mathbf{u} = x_{\mathbf{u}}\mathbf{a} + y_{\mathbf{u}}\mathbf{b}, \quad \mathbf{v} = x_{\mathbf{v}}\mathbf{a} + y_{\mathbf{v}}\mathbf{b}.$$

The equalities

$$\xi_{\mathbf{u}} = x_{\mathbf{u}}\xi_{\mathbf{a}} + y_{\mathbf{u}}\xi_{\mathbf{b}}, \quad \eta_{\mathbf{u}} = x_{\mathbf{u}}\eta_{\mathbf{a}} + y_{\mathbf{u}}\eta_{\mathbf{b}},$$
$$\xi_{\mathbf{v}} = x_{\mathbf{v}}\xi_{\mathbf{a}} + y_{\mathbf{v}}\xi_{\mathbf{b}}, \quad \eta_{\mathbf{v}} = x_{\mathbf{v}}\eta_{\mathbf{a}} + y_{\mathbf{v}}\eta_{\mathbf{b}},$$

which are implicit in 5·31, give

$$\xi_{\mathbf{u}}\eta_{\mathbf{v}} - \eta_{\mathbf{u}}\xi_{\mathbf{v}} = (x_{\mathbf{u}}y_{\mathbf{v}} - y_{\mathbf{u}}x_{\mathbf{v}})(\xi_{\mathbf{a}}\eta_{\mathbf{b}} - \eta_{\mathbf{a}}\xi_{\mathbf{b}}),$$

and similarly each of the other five expressions whose squares appear on the right of 5·21 has its value for $\mathbf{u}$ and $\mathbf{v}$ given by multiplying its value for $\mathbf{a}$ and $\mathbf{b}$ by $x_{\mathbf{u}}y_{\mathbf{v}} - y_{\mathbf{u}}x_{\mathbf{v}}$, a constant that depends only on the four vectors $\mathbf{a}, \mathbf{b}, \mathbf{u}, \mathbf{v}$. But if $\delta_{\mathbf{uv}}$ is any angle between $\mathbf{u}$ and $\mathbf{v}$, the expression on the left of 5·21 has by definition the value $1 - \cos^2 \delta_{\mathbf{uv}}$, and therefore each side of 5·21 has the value $\sin^2 \delta_{\mathbf{uv}}$. Thus 5·21 and 5·31 imply that identically

$$\sin^2 \delta_{\mathbf{uv}} = (x_{\mathbf{u}}y_{\mathbf{v}} - y_{\mathbf{u}}x_{\mathbf{v}})^2 \sin^2 \delta_{\mathbf{ab}},$$

and if we have chosen a value of $\delta_{\mathbf{ab}}$, we can distinguish rationally those angles between $\mathbf{u}$ and $\mathbf{v}$ whose sine is $(x_{\mathbf{u}}y_{\mathbf{v}} - y_{\mathbf{u}}x_{\mathbf{v}}) \sin \delta_{\mathbf{ab}}$ from those whose sine is the negative of this. We use $\epsilon_{\mathbf{ab}}$ for a chosen value of $\delta_{\mathbf{ab}}$, and $\epsilon_{\mathbf{uv}}$ for an angle between $\mathbf{u}$ and $\mathbf{v}$ such that

5·32 $$\sin \epsilon_{\mathbf{uv}} = (x_{\mathbf{u}}y_{\mathbf{v}} - y_{\mathbf{u}}x_{\mathbf{v}}) \sin \epsilon_{\mathbf{ab}}.$$

To interchange $\mathbf{u}$ and $\mathbf{v}$ is to reverse the sign of the factor $x_{\mathbf{u}}y_{\mathbf{v}} - y_{\mathbf{u}}x_{\mathbf{v}}$; hence

5·33 $$\sin \epsilon_{\mathbf{vu}} = - \sin \epsilon_{\mathbf{uv}}.$$

To interchange $\mathbf{a}$ and $\mathbf{b}$ is to reverse the parts played by x and y, and in this way to change the sign of $x_{\mathbf{u}}y_{\mathbf{v}} - y_{\mathbf{u}}x_{\mathbf{v}}$, and therefore if $\sin \epsilon_{\mathbf{uv}}$ is to be unaltered, $\sin \epsilon_{\mathbf{ba}}$ must denote $-\sin \epsilon_{\mathbf{ab}}$, as is otherwise evident since from their definitions

5·34 $$x_{\mathbf{a}} = 1, \quad y_{\mathbf{a}} = 0, \quad x_{\mathbf{b}} = 0, \quad y_{\mathbf{b}} = 1.$$

Thus the order in which the directions are taken is essential if $\epsilon_{\mathbf{ab}}$ and $\epsilon_{\mathbf{uv}}$ are to have significance, and we say that $\epsilon_{\mathbf{uv}}$ is an angle *from* $\mathbf{u}$ *to* $\mathbf{v}$ *conformable* with the angle $\epsilon_{\mathbf{ab}}$ *from* $\mathbf{a}$ *to* $\mathbf{b}$.

Ex. xxi. The vectors

$$(-1, 2, 4, 2),\ (2, 2, 0, -1),\ (31, 58, 36, -2),$$
$$(31-42\sqrt{3},\ 58-6\sqrt{3},\ 36+48\sqrt{3},\ -2+39\sqrt{3})$$

are coplanar vectors of which 5, -3, -75, 150 are measures. The first two are perpendicular; the cosine of angles between the directions in which the last two have the measures given is $-\frac{1}{2}$. If we denote the four directions by $\mathbf{k}$, $\mathbf{l}$, $\mathbf{m}$, $\mathbf{n}$, the vectors are $5\mathbf{k}$, $-3\mathbf{l}$, $-75\mathbf{m}$, $150\mathbf{n}$, and identically

$$-75\mathbf{m}=9.5\mathbf{k}-20.3\mathbf{l},\quad 150\mathbf{n}=(9+12\sqrt{3}).5\mathbf{k}-(20-15\sqrt{3}).3\mathbf{l}.$$

Hence with conformable measurement $\sin\epsilon_{\mathbf{mn}}$ is

$$-[\{3(3\sqrt{3}-4)+4(3+4\sqrt{3})\}/50]\sin\epsilon_{\mathbf{kl}},$$

and the angles conformable with an angle of $\frac{1}{2}\pi$ from $\mathbf{k}$ to $\mathbf{l}$ are the angles with cosine $-\frac{1}{2}$ and sine $-\frac{1}{2}\sqrt{3}$, that is, are the angles $(2n-\frac{2}{3})\pi$.

The importance of measuring angles conformably when the directions concerned are coplanar is due to two propositions which we proceed to prove.

Suppose the unit vectors $\mathbf{u}$, $\mathbf{v}$, coplanar with $\mathbf{a}$ and $\mathbf{b}$ and given by 5·31, not to be collinear, and let $\mathbf{p}$, $\mathbf{q}$ be any two unit vectors in the vecplane that contains the four vectors $\mathbf{a}$, $\mathbf{b}$, $\mathbf{u}$, $\mathbf{v}$. Then $\mathbf{p}$, $\mathbf{q}$ can be expressed either in terms of $\mathbf{a}$ and $\mathbf{b}$, by the formulae

5·35 $$\mathbf{p}=x_{\mathbf{p}}\mathbf{a}+y_{\mathbf{p}}\mathbf{b},\quad \mathbf{q}=x_{\mathbf{q}}\mathbf{a}+y_{\mathbf{q}}\mathbf{b},$$

or in terms of $\mathbf{u}$ and $\mathbf{v}$ by formulae in which we will use a temporary notation, writing

5·36 $$\mathbf{p}=f_{\mathbf{p}}\mathbf{u}+g_{\mathbf{p}}\mathbf{v},\quad \mathbf{q}=f_{\mathbf{q}}\mathbf{u}+g_{\mathbf{q}}\mathbf{v}.$$

If in 5·36 we substitute for $\mathbf{u}$, $\mathbf{v}$ from 5·31, we have

$$\mathbf{p}=(f_{\mathbf{p}}x_{\mathbf{u}}+g_{\mathbf{p}}x_{\mathbf{v}})\mathbf{a}+(f_{\mathbf{p}}y_{\mathbf{u}}+g_{\mathbf{p}}y_{\mathbf{v}})\mathbf{b},$$
$$\mathbf{q}=(f_{\mathbf{q}}x_{\mathbf{u}}+g_{\mathbf{q}}x_{\mathbf{v}})\mathbf{a}+(f_{\mathbf{q}}y_{\mathbf{u}}+g_{\mathbf{q}}y_{\mathbf{v}})\mathbf{b},$$

whence by comparison with 5·35

$$x_{\mathbf{p}}=f_{\mathbf{p}}x_{\mathbf{u}}+g_{\mathbf{p}}x_{\mathbf{v}},\quad y_{\mathbf{p}}=f_{\mathbf{p}}y_{\mathbf{u}}+g_{\mathbf{p}}y_{\mathbf{v}},$$
$$x_{\mathbf{q}}=f_{\mathbf{q}}x_{\mathbf{u}}+g_{\mathbf{q}}x_{\mathbf{v}},\quad y_{\mathbf{q}}=f_{\mathbf{q}}y_{\mathbf{u}}+g_{\mathbf{q}}y_{\mathbf{v}},$$

and therefore identically

$$x_{\mathbf{p}}y_{\mathbf{q}}-y_{\mathbf{p}}x_{\mathbf{q}}=(f_{\mathbf{p}}g_{\mathbf{q}}-g_{\mathbf{p}}f_{\mathbf{q}})(x_{\mathbf{u}}y_{\mathbf{v}}-y_{\mathbf{u}}x_{\mathbf{v}});$$

hence from 5·32

$$(x_{\mathbf{p}}y_{\mathbf{q}}-y_{\mathbf{p}}x_{\mathbf{q}})\sin\epsilon_{\mathbf{ab}}=(f_{\mathbf{p}}g_{\mathbf{q}}-g_{\mathbf{p}}f_{\mathbf{q}})\sin\epsilon_{\mathbf{uv}}.$$

On the left we have the value of $\sin\epsilon_{\mathbf{pq}}$ determined conformably with $\epsilon_{\mathbf{ab}}$, and on the right the value of $\sin \epsilon_{\mathbf{pq}}$ determined conformably with $\epsilon_{\mathbf{uv}}$, and these values have been proved equal on the assumption that $\epsilon_{\mathbf{ab}}$ and $\epsilon_{\mathbf{uv}}$ are conformable with each other: two coplanar angles which are both conformable with a third angle coplanar with them are conformable with each other.

Again, let **u** be any unit vector coplanar with two unit vectors **a**, **b** that are not collinear, and expressible therefore as $x_{\mathbf{u}}\,\mathbf{a} + y_{\mathbf{u}}\,\mathbf{b}$. Identically,

$$(x_{\mathbf{u}}\xi_{\mathbf{a}} + y_{\mathbf{u}}\xi_{\mathbf{b}})^2 + (x_{\mathbf{u}}\eta_{\mathbf{a}} + y_{\mathbf{u}}\eta_{\mathbf{b}})^2 + (x_{\mathbf{u}}\zeta_{\mathbf{a}} + y_{\mathbf{u}}\zeta_{\mathbf{b}})^2 + (x_{\mathbf{u}}\tau_{\mathbf{a}} + y_{\mathbf{u}}\tau_{\mathbf{b}})^2$$
$$= x_{\mathbf{u}}^2(\xi_{\mathbf{a}}^2 + \eta_{\mathbf{a}}^2 + \zeta_{\mathbf{a}}^2 + \tau_{\mathbf{a}}^2) + 2x_{\mathbf{u}}y_{\mathbf{u}}(\xi_{\mathbf{a}}\xi_{\mathbf{b}} + \eta_{\mathbf{a}}\eta_{\mathbf{b}} + \zeta_{\mathbf{a}}\zeta_{\mathbf{b}} + \tau_{\mathbf{a}}\tau_{\mathbf{b}})$$
$$+ y_{\mathbf{u}}^2(\xi_{\mathbf{b}}^2 + \eta_{\mathbf{b}}^2 + \zeta_{\mathbf{b}}^2 + \tau_{\mathbf{b}}^2),$$

and therefore since the vectors are unit vectors

5·37 $$x_{\mathbf{u}}^2 + 2x_{\mathbf{u}}y_{\mathbf{u}}\cos\epsilon_{\mathbf{ab}} + y_{\mathbf{u}}^2 = 1.$$

Also

$$\cos\epsilon_{\mathbf{au}} = \xi_{\mathbf{a}}(x_{\mathbf{u}}\xi_{\mathbf{a}} + y_{\mathbf{u}}\xi_{\mathbf{b}}) + \eta_{\mathbf{a}}(x_{\mathbf{u}}\eta_{\mathbf{a}} + y_{\mathbf{u}}\eta_{\mathbf{b}}) + \zeta_{\mathbf{a}}(x_{\mathbf{u}}\zeta_{\mathbf{a}} + y_{\mathbf{u}}\zeta_{\mathbf{b}})$$
$$+ \tau_{\mathbf{a}}(x_{\mathbf{u}}\tau_{\mathbf{a}} + y_{\mathbf{u}}\tau_{\mathbf{b}}),$$

that is, $$\cos\epsilon_{\mathbf{au}} = x_{\mathbf{u}} + y_{\mathbf{u}}\cos\epsilon_{\mathbf{ab}},$$

and similarly $$\cos\epsilon_{\mathbf{ub}} = x_{\mathbf{u}}\cos\epsilon_{\mathbf{ab}} + y_{\mathbf{u}},$$

while from 5·34

5·38 $$\sin\epsilon_{\mathbf{au}} = y_{\mathbf{u}}\sin\epsilon_{\mathbf{ab}}, \quad \sin\epsilon_{\mathbf{ub}} = x_{\mathbf{u}}\sin\epsilon_{\mathbf{ab}}.$$

Hence identically

$$\cos\epsilon_{\mathbf{au}}\cos\epsilon_{\mathbf{ub}} - \sin\epsilon_{\mathbf{au}}\sin\epsilon_{\mathbf{ub}} = (x_{\mathbf{u}}^2 + y_{\mathbf{u}}^2)\cos\epsilon_{\mathbf{ab}}$$
$$+ x_{\mathbf{u}}y_{\mathbf{u}}(1 + \cos^2\epsilon_{\mathbf{ab}} - \sin^2\epsilon_{\mathbf{ab}})$$
$$= (x_{\mathbf{u}}^2 + 2x_{\mathbf{u}}y_{\mathbf{u}}\cos\epsilon_{\mathbf{ab}} + y_{\mathbf{u}}^2)\cos\epsilon_{\mathbf{ab}},$$
$$\sin\epsilon_{\mathbf{au}}\cos\epsilon_{\mathbf{ub}} + \cos\epsilon_{\mathbf{au}}\sin\epsilon_{\mathbf{ub}} = (x_{\mathbf{u}}^2 + 2x_{\mathbf{u}}y_{\mathbf{u}}\cos\epsilon_{\mathbf{ab}} + y_{\mathbf{u}}^2)\sin\epsilon_{\mathbf{ab}},$$

and therefore from 5·37,

5·39 $$\cos(\epsilon_{\mathbf{au}} + \epsilon_{\mathbf{ub}}) = \cos\epsilon_{\mathbf{ab}}, \quad \sin(\epsilon_{\mathbf{au}} + \epsilon_{\mathbf{ub}}) = \sin\epsilon_{\mathbf{ab}}.$$

These formulae express the second fundamental property of angles measured conformably: if three directions are coplanar, the sum of conformable angles from the first to the second and from the second to the third is an angle from the first to the third conformable with the other two angles.

5·4. To find the measures of a vector, or the angles between two vectors, in terms of coefficients in a vecframe **abcd**, we have only to substitute for the absolute components from the formulae typified by 3·54. We see at once that if r, s are measures of two vectors $\mathbf{r}$, $\mathbf{s}$, and $\delta_{\mathbf{rs}}$ is an angle between the directions in which they are measured, then

$$5{\cdot}41 \quad rs\cos\delta_{\mathbf{rs}} = a^2 x_{\mathbf{r}} x_{\mathbf{s}} + b^2 y_{\mathbf{r}} y_{\mathbf{s}} + c^2 z_{\mathbf{r}} z_{\mathbf{s}} + d^2 t_{\mathbf{r}} t_{\mathbf{s}} + ab\,(x_{\mathbf{r}} y_{\mathbf{s}} + y_{\mathbf{r}} x_{\mathbf{s}}) \cos\delta_{\mathbf{ab}} + ac\,(x_{\mathbf{r}} z_{\mathbf{s}} + z_{\mathbf{r}} x_{\mathbf{s}}) \cos\delta_{\mathbf{ac}} + ad\,(x_{\mathbf{r}} t_{\mathbf{s}} + t_{\mathbf{r}} x_{\mathbf{s}}) \cos\delta_{\mathbf{ad}} + bc\,(y_{\mathbf{r}} z_{\mathbf{s}} + z_{\mathbf{r}} y_{\mathbf{s}}) \cos\delta_{\mathbf{bc}} + bd\,(y_{\mathbf{r}} t_{\mathbf{s}} + t_{\mathbf{r}} y_{\mathbf{s}}) \cos\delta_{\mathbf{bd}} + cd\,(z_{\mathbf{r}} t_{\mathbf{s}} + t_{\mathbf{r}} z_{\mathbf{s}}) \cos\delta_{\mathbf{cd}},$$

where a, b, c, d are measures of $\mathbf{a}$, $\mathbf{b}$, $\mathbf{c}$, $\mathbf{d}$; the square of r is found by making $\mathbf{s}$ coincide with $\mathbf{r}$. The expression for $rs\cos\delta_{\mathbf{rs}}$ can be simplified radically in two ways, which are independent, by a restriction on the vectors of reference.

If the vectors are unit vectors, their measures no longer appear explicitly in the formula, and the only constants that occur are the six cosines. The vecframe is then said to be Cartesian. A Cartesian vecframe is determined wholly by four directions, and any* four directions that are not cospatial, taken in a definite order, provide one definite Cartesian vecframe.

The use of a Cartesian vecframe does not necessarily diminish the number of terms involved in 5·41, but if two of the vectors of reference are perpendicular there is an immediate reduction. If p, q are any two vecplanes that are perpendicular to each other, and if $\mathbf{a}$, $\mathbf{b}$ are perpendicular proper vectors in p and $\mathbf{c}$, $\mathbf{d}$ perpendicular proper vectors in q, then any two of the four vectors are perpendicular, and 5·41 becomes

$$5{\cdot}42 \qquad rs\cos\delta_{\mathbf{rs}} = a^2 x_{\mathbf{r}} x_{\mathbf{s}} + b^2 y_{\mathbf{r}} y_{\mathbf{s}} + c^2 z_{\mathbf{r}} z_{\mathbf{s}} + d^2 t_{\mathbf{r}} t_{\mathbf{s}}:$$

the vecframe is said to be rectangular.

If the vecframe is both Cartesian and rectangular, the formula takes the simplest† form that is possible, namely

$$5{\cdot}43 \qquad rs\cos\delta_{\mathbf{rs}} = x_{\mathbf{r}} x_{\mathbf{s}} + y_{\mathbf{r}} y_{\mathbf{s}} + z_{\mathbf{r}} z_{\mathbf{s}} + t_{\mathbf{r}} t_{\mathbf{s}};$$

the corresponding formula giving measures is

$$5{\cdot}44 \qquad r^2 = x_{\mathbf{r}}^2 + y_{\mathbf{r}}^2 + z_{\mathbf{r}}^2 + t_{\mathbf{r}}^2.$$

Ex. xxii. The sets of vectors in Example xiii (p. 17) form rectangular vecframes. The measures in the first set are ± 15, ± 15, ± 5, ± 15, and those in

* In complex space this statement needs modification: if a self-perpendicular direction is used, the frame can not be Cartesian. This is, I think, why the logic of frames with isotropic axes has remained obscure.

† The simplest in real space; in complex space, we can obtain

$$rs\cos\delta_{\mathbf{rs}} = x_{\mathbf{r}} y_{\mathbf{s}} + y_{\mathbf{r}} x_{\mathbf{s}} + z_{\mathbf{r}} t_{\mathbf{s}} + t_{\mathbf{r}} z_{\mathbf{s}},$$

which is not merely as simple as 5·43 but leads to a formula for r^2 that is simpler than 5·44, namely, $r^2 = 2x_{\mathbf{r}} y_{\mathbf{r}} + 2z_{\mathbf{r}} t_{\mathbf{r}}$.

the second set are ± 3, ± 5, ± 15, ± 3, and by dividing each vector by one of its measures we obtain two vecframes that are Cartesian as well as rectangular.

The amount of freedom in the choice of a rectangular vecframe is readily estimated. The vector **a** may be arbitrary, subject only to being proper; thus the four absolute components of **a** are arbitrary. Then **b** may be any proper vector in the vecspace perpendicular to **a**; if this vecspace is built on three vectors **f**, **g**, **h**, there are three arbitrary coefficients in the expression of **b**. The vecplane perpendicular to **ab** is determinate, and **c**, an arbitrary member of this vecplane, involves two coefficients. Finally **d**, belonging to the vecline perpendicular to **abc**, depends on one parameter. Thus there are in all ten constants, arbitrary but for the condition that the vectors determined are to be proper. We can reach the same conclusion more rapidly but with less confidence by mere enumeration; there are at our disposal sixteen constants, the absolute components of four vectors, and these constants must satisfy six equations, the equations expressing that six cosines vanish; it is natural to find that the difference between the number of conditions and the number of constants measures the freedom of our choice.

If the vecframe is to be Cartesian as well as rectangular, the vectors must be divided by four constants that are determinate in value but ambiguous in sign. Thus a general set of formulae for the determination of rectangular Cartesian vecframes contains six independent constants and four disposable signs. A general theory of sets of formulae of this kind, not restricted to four-dimensional space, was discovered by Cayley. The results for four-dimensional space, first given by Euler, are easily verified; in terms of twelve constants denoted by a_{rs}, where r, s take *different* values from 1 to 4 and

5·45 $$a_{sr} = -a_{rs},$$

they may be written in the form

5·46
$$\begin{aligned}
\pm D\mathbf{a} = \{&p-(a_{12}^2+a_{13}^2+a_{14}^2),\quad a_{12}-(a_{13}a_{23}+a_{14}a_{24})+a_{34}k,\\
&a_{13}-(a_{14}a_{34}+a_{12}a_{32})+a_{42}k,\quad a_{14}-(a_{12}a_{42}+a_{13}a_{43})+a_{23}k\},\\
\pm D\mathbf{b} = \{&a_{21}-(a_{24}a_{14}+a_{23}a_{13})+a_{43}k,\quad p-(a_{21}^2+a_{23}^2+a_{24}^2),\\
&a_{23}-(a_{21}a_{31}+a_{24}a_{34})+a_{14}k,\quad a_{24}-(a_{23}a_{43}+a_{21}a_{41})+a_{31}k\},\\
\pm D\mathbf{c} = \{&a_{31}-(a_{32}a_{12}+a_{34}a_{14})+a_{24}k,\quad a_{32}-(a_{34}a_{24}+a_{31}a_{21})+a_{41}k,\\
&p-(a_{31}^2+a_{32}^2+a_{34}^2),\quad a_{34}-(a_{31}a_{41}+a_{32}a_{42})+a_{12}k\},\\
\pm D\mathbf{d} = \{&a_{41}-(a_{43}a_{13}+a_{42}a_{12})+a_{32}k,\quad a_{42}-(a_{41}a_{21}+a_{43}a_{23})+a_{13}k,\\
&a_{43}-(a_{42}a_{32}+a_{41}a_{31})+a_{21}k,\quad p-(a_{41}^2+a_{42}^2+a_{43}^2)\},
\end{aligned}$$

where

5·47 $$k = a_{12}a_{34} + a_{13}a_{42} + a_{14}a_{23},$$

5·48 $$p = \tfrac{1}{2}(1 + a_{12}^2 + a_{34}^2 + a_{13}^2 + a_{42}^2 + a_{14}^2 + a_{23}^2 - k^2),$$

5·49 $$D = p + k^2,$$

and the ambiguities of sign are independent of each other.

We can of course dispense with six of the constants by means of 5·45, and use for example only the six in terms of which k is expressed, but only at the cost of rendering the scheme of vectors quite unintelligible. As it is, the secret lies in associating with a_{pq} a conjugate a_{rs}, where p, q, r, s are the four numbers 1, 2, 3, 4 *taken in an order that can be reached by an even number of transpositions from the primitive arrangement.* If a_{pq} and a_{rs} are conjugate, so are a_{qp} and a_{sr}, and the products $a_{pq}a_{rs}$, $a_{qp}a_{sr}$ have the same value; thus k, although expressed in terms of a particular selection of the constants, does not really depend on the selection made.

6. LINES, PLANES, AND SPACES

6·1. We can now resume definitions connected with four-dimensional space, not with vectors. By a direction of a step is meant a direction of the vector of the step, and the *length* of a step in one of its directions is the measure of its vector in that direction; a length of a step is called also a *distance* from the beginning to the end of the step. A zero step has every direction, but zero is its only length. A proper step has two directions, one opposite to the other, and its length in one direction is the negative of its length in the other.

A *measured* proper step has a definite direction, and there are angles between measured proper steps. If a number of proper steps have coplanar vectors, the measurement of angles between them can be made conformable.

It is important to observe that all the formulae of plane trigonometry are preserved by the definitions of lengths and angles. If ABC is a triangle, that is, a set of three points, and if $\mathbf{a}, \mathbf{b}, \mathbf{c}$ are the vectors of the steps BC, CA, AB and have measures a, b, c, then since $-\mathbf{a}$ is $\mathbf{b} + \mathbf{c}$,

$$\xi\mathbf{a}^2 + \eta\mathbf{a}^2 + \zeta\mathbf{a}^2 + \tau\mathbf{a}^2 = (\xi\mathbf{b} + \xi\mathbf{c})^2 + (\eta\mathbf{b} + \eta\mathbf{c})^2 + (\zeta\mathbf{b} + \zeta\mathbf{c})^2 + (\tau\mathbf{b} + \tau\mathbf{c})^2,$$

that is,

6·11 $$a^2 = b^2 + 2bc \cos \epsilon_{\mathbf{bc}} + c^2;$$

the angle $\epsilon_{\mathbf{bc}}$ is an angle between a direction of CA and a direction of AB, and is therefore an *external* angle of the triangle and is denoted in the usual notation by $\pi - A$. Thus 6·11 takes the familiar form

6·12 $$a^2 = b^2 - 2bc \cos A + c^2.$$

Again, the unit vectors in the directions in which $\mathbf{a}$, $\mathbf{b}$, $\mathbf{c}$ have measures a, b, c are $\mathbf{a}/a$, $\mathbf{b}/b$, $\mathbf{c}/c$, and in terms of these vectors the relation

$$-\mathbf{a} = \mathbf{b} + \mathbf{c}$$

becomes

$$(\mathbf{a}/a) = (-b/a)(\mathbf{b}/b) + (-c/a)(\mathbf{c}/c),$$

and the formulae

$$\sin \epsilon_{\mathbf{au}} = y_{\mathbf{u}} \sin \epsilon_{\mathbf{ab}}, \quad \sin \epsilon_{\mathbf{ub}} = x_{\mathbf{u}} \sin \epsilon_{\mathbf{ab}}$$

of 5·3, with the substitution of $\mathbf{a}/a$, $\mathbf{b}/b$, $\mathbf{c}/c$ for $\mathbf{u}$, $\mathbf{a}$, $\mathbf{b}$, are equivalent to

6·13 $$\sin \epsilon_{\mathbf{ba}} = -(c/a) \sin \epsilon_{\mathbf{bc}}, \quad \sin \epsilon_{\mathbf{ac}} = -(b/a) \sin \epsilon_{\mathbf{bc}},$$

that is, to

6·14 $$\sin A/a = \sin B/b = \sin C/c.$$

6·2. By the *line, plane,* or *space,* through a given point Q with a given vecline, vecplane, or vecspace, is meant the class to which a point R must belong if the vector of the step QR is to belong to the given vecline, vecplane, or vecspace.

Although it is necessary to define a line, plane, or space as a line, plane, or space through some one point, it is a simple matter to prove that this point does not occupy a special position in the line, plane, or space, and that in fact the class of points does not bear any relation to the point used in the definition that it does not bear to every one of its members; it will be sufficient if we give the proof for one case only, and we deal with the case of a plane. Let P, R be any two points in the plane through a given point Q with the vecplane p. By definition, the vector $P - Q$ belongs to p, and therefore so also does $Q - P$; also by definition $R - Q$ belongs to p, and therefore $(R - Q) + (Q - P)$, which is $R - P$, also belongs to p; thus every point in the plane through Q with vecplane p belongs to the plane through P with the same vecplane. Conversely, if S is in the plane through P with vecplane p, the vectors $S - P$, $P - Q$ belong to p and therefore so does $S - Q$; that is, every point in the plane through P with vecplane p is in the plane through Q with this vecplane. Combining the two results, we find that if P is any point of the plane through Q with a given vecplane, the plane through P with this vecplane coincides with the plane through Q.

Ex. xxiii. The line through the point $(3, -1, 5, 1)$ with the vecline built on the vector $(2, 1, -2, -1)$ is the class of all tetrads of the form

$$(3+2x,\ -1+x,\ 5-2x,\ 1-x).$$

One point of this line is $(5, 0, 3, 0)$, and the typical point may be presented also in the form $(5+2X,\ X,\ 3-2X,\ -X)$.

Ex. xxiv. The plane through $(3, -1, 5, 1)$ with the vecplane discussed in Example vii (p. 12) is the class of points of the form

$$(3+2x+y,\ -1+x-y,\ 5-2x-y,\ 1-x+3y).$$

The point $(8, 0, 0, 2)$ belongs to the plane, and if the plane is referred to this point while the vecplane is regarded as built on $(-1, -2, 1, 4)$ and $(3, 0, -3, 2)$, the typical point takes the form $(8-X+3Y,\ -2X,\ X-3Y,\ 2+4X+2Y)$. To verify that the plane itself is not affected by the manner in which the typical point is expressed, we have only to remark that the whole set of equations

$$3+2x+y=8-X+3Y, \quad -1+x-y=-2X,$$
$$5-2x-y=X-3Y, \quad 1-x+3y=2+4X+2Y$$

is equivalent both to

$$X=\tfrac{1}{2}-\tfrac{1}{2}x+\tfrac{1}{2}y, \quad Y=-\tfrac{3}{2}+\tfrac{1}{2}x+\tfrac{1}{2}y$$

and to

$$x=2-X+Y, \quad y=1+X+Y.$$

If P, Q are any two distinct points, there is one and only one line which contains both P and Q, for the vecline is determined by the vector of the step PQ.

If P, Q, R are three points that are not collinear, the vectors of the steps PQ, PR are not collinear, and there is one and only one vecplane which contains these vectors; the plane through P with this vecplane contains Q and R and is the only plane through P that does contain both of these points. That is, if three points are not collinear, there is one and only one plane that contains them all.

A similar proof shews that if four points are not coplanar there is one and only one space to which they all belong.

From these results it follows that there is one and only one plane that includes a given line and contains a given point outside the line, and one and only one space that includes a given plane and contains a given point outside the plane, and also that if there is no plane that includes two given lines, then there is one and only one space that includes them both.

6·3. Two lines, two planes, or two spaces are said to be *parallel* if they have the same vecline, vecplane, or vecspace. A line is described as parallel to a plane or to a space if the vecline of the one is included in the vecplane or vecspace of the other, and a plane is said to be parallel to a space if its vecplane is part of the corresponding vecspace.

It follows at once from the definitions that parallel lines, parallel planes, or parallel spaces that have a single point in common coincide entirely, that a line parallel to a plane or a space either forms part of the plane or the space or has no point in common with it, and that if a plane and a space are parallel either the plane is in the space or the two have no point of intersection.

Let P be any point in a space whose vecspace is built on three vectors $\mathbf{a}$, $\mathbf{b}$, $\mathbf{c}$, and let Q be any point on a line whose vecline contains a proper vector $\mathbf{d}$; let $\mathbf{p}$ be the vector of the step PQ. Then between the five vectors $\mathbf{a}$, $\mathbf{b}$, $\mathbf{c}$, $\mathbf{d}$, $\mathbf{p}$ there is an effective linear relation

6·31 $$a\mathbf{a} + b\mathbf{b} + c\mathbf{c} + d\mathbf{d} + p\mathbf{p} = \mathbf{0}.$$

If in this equation p is zero, the equation expresses that $\mathbf{a}$, $\mathbf{b}$, $\mathbf{c}$, $\mathbf{d}$ are cospatial, that is, that the line through Q is parallel to the space through P; in this case the line either forms part of the space or has no point in common with the space. If p is not zero, 6·31 can be written as

6·32 $$\mathbf{p} + (d/p)\,\mathbf{d} = (-a/p)\,\mathbf{a} + (-b/p)\,\mathbf{b} + (-c/p)\,\mathbf{c},$$

and shews that if R is the point such that QR has the vector $(d/p)\,\mathbf{d}$, which is necessarily a point in the given line through Q, then the vector of PR belongs to the vecspace $\mathbf{abc}$ and therefore R is in the given space through P; that is, there is at least one point common to the line and the space. If *two* points of a line belong to a space, the whole line is included in the space and is therefore parallel to the space. Hence if a line and a space are not parallel, they have one and only one common point.

Ex. xxv. If P, Q, R denote the points

$$(3, -1, -1, -6),\ (6, 2, 0, -3),\ (-6, -7, 8, -6)$$

and $\mathbf{a}$, $\mathbf{b}$, $\mathbf{c}$ the vectors $(1, -1, -1, 3)$, $(7, 2, -7, 4)$, $(1, 2, 3, 4)$, the line through P whose vecline contains $(3, 2, -3, 0)$ cuts the space through Q with vecspace $\mathbf{abc}$ in the point R, for the vector $R-P$ is $(-9, -6, 9, 0)$, a multiple of $(3, 2, -3, 0)$, and the vector $R-Q$ is $(-12, -9, 8, -3)$ and is expressible as $3\mathbf{a}-2\mathbf{b}-\mathbf{c}$; also the line through P whose vecline contains $(3, 3, 1, 5)$ cuts the same space in $(-3, -7, -3, -16)$, for if this point is denoted by S, the vector $S-P$ is $-2(3, 3, 1, 5)$ and the vector $S-Q$ is $\mathbf{a}-\mathbf{b}-3\mathbf{c}$.

If a plane and a space are not parallel, there are lines in the plane that are not parallel to the space, and it is easy to deduce from the last result that the points common to the plane and the space compose a definite line. An extension of the same argument shews that the common points of two spaces that are not parallel are the points of some one plane. To exhibit a given plane as the intersection of two spaces, all that is necessary is to take a point outside the plane, to construct the space that includes the plane and contains this point, and then to take a point outside this space and to repeat the construction with the plane and the second point.

Ex. xxvi. With the notation of the last example, the plane through P with vecplane built on $(3, 2, -3, 0)$ and $(3, 3, 1, 5)$ cuts the space through Q with vecspace $\mathbf{abc}$ in the line RS.

Since in general there is no vecspace that includes two given vecplanes, it is exceptional for two given planes to be in one space. If two planes p, q are in one space, and if q is represented as the intersection of this space with a second space, either p cuts this second space in a line or p is parallel to the second space; in the first case the line, since it is common to the two spaces, is part of q; in the second case p is parallel to q, since the vecplane of p must be common to the vecspaces of the two spaces. Hence if two planes are in one space, either they are parallel or they intersect in a line. Conversely, if two planes p, q are parallel and distinct, and if P is a point of one and Q a point of the other, the space through P whose vecspace includes the vecplane of p and q and contains the vector of the step PQ is a space that includes both p and q; and if two planes p, q have a common line of which P, Q are distinct points, then if R is a point of p not in PQ and S a point of q not in PQ, the space $PQRS$ includes the two planes. Thus if two planes are parallel or intersect in a line, there is a space that includes them both.

Ex. xxvii. If F, G denote the points $(-2, 1, -2, 4)$, $(6, 4, -10, 5)$, and if $\mathbf{k}$, $\mathbf{l}$, $\mathbf{m}$, $\mathbf{n}$ are the vectors so denoted in Example xi (p. 15), and shewn there to be cospatial, the vector $G-F$ is cospatial with these four. The plane through F with vecplane $\mathbf{kl}$ and the plane through G with vecplane $\mathbf{mn}$ are in one space, and they have in common the two points $(-2, 4, -2, -3)$, $(8, 0, -12, 13)$, for if these points are U, V, the vectors $U-F$, $U-G$ are $\mathbf{k}-2\mathbf{l}$, $-2\mathbf{m}+2\mathbf{n}$, and the vectors $V-F$, $V-G$ are $3\mathbf{k}+4\mathbf{l}$, $2\mathbf{m}-4\mathbf{n}$. The whole of the line UV is common to the two planes, and the vector $V-U$ is $(10, -4, -10, 16)$, which is expressible both as $2\mathbf{k}+6\mathbf{l}$ and as $4\mathbf{m}-6\mathbf{n}$.

To discover the nature of the intersection of two planes p, q when there is no space that includes them both, we may again regard q as the intersection of two spaces; since neither of these spaces can include p or be parallel to p, the plane p cuts the first space in a definite line and this line cuts the second space in a definite point, which, being in both spaces, is a point of q. That is. if there is no space in which two given planes are both included, the planes have one and only one common point.

Ex. xxviii. If P, Q, $\mathbf{a}$, $\mathbf{b}$, $\mathbf{c}$ have the same meanings as in Example xxv (p. 33), and if $\mathbf{d}$ denotes the vector $(3, 2, -3, 0)$, the identity used in Example xii (p. 16) may be written

$$(Q-P)+3\mathbf{a}-2\mathbf{b}-\mathbf{c}+3\mathbf{d}=0,$$

and shews that the plane through P with vecplane $\mathbf{ab}$ and the plane through Q with vecplane $\mathbf{cd}$ intersect in a point T which is such that $T-P$ is $-3\mathbf{a}+2\mathbf{b}$ and $T-Q$ is $-\mathbf{c}+3\mathbf{d}$; this point is $(14, 6, -12, -7)$.

If a line and a plane have two points in common, the line is part of the plane. If a line and a plane have one point in common or are parallel, there is a space that includes them both. On the other hand, if a line and a plane are included in one space, a second space can be constructed to include the plane; if the line is parallel to this second space, it is parallel to both spaces and therefore parallel to the plane, while if the line is not parallel to the second space it cuts this space in a definite point which belongs to both spaces and therefore belongs to the plane: thus if a line and a plane are included in one space, either they are parallel or they have a common point.

If two lines either intersect or are parallel, there is a plane that includes them both, and a modification of arguments used above shews that the converse is true: if two lines are included in one plane, either they are parallel or they intersect.

By means of intersections we can define certain types of distance. Given a point P, a space, and a direction $\mathbf{u}$, there is one and only one line through P with the given direction $\mathbf{u}$, and if this line is not parallel to the space it cuts the space in one definite point Q. By the distance from *P to the space* in the direction $\mathbf{u}$ is meant the distance from P to Q in that direction. The distance is zero, whatever the direction, if and only if P is in the space. If $\mathbf{u}$ is one of the two directions perpendicular to the space, the distance is a perpendicular distance, or a length of the perpendicular from P to the space; since parallelism is incompatible with perpendicularity, perpendicular distances always* exist.

Similarly, distances between a point P and a given plane q can be defined by means of the intersection of q with the plane through P having a given vecplane, and distances between a point P and a line Q by means of the intersection of Q with the space through P having a given vecspace, and in each case there is one pair of perpendicular distances. Only in these cases the *directions* in which the distances are to be measured are not assignable in advance.

6·4. If $\mathbf{n}$ is a proper vector perpendicular to a given vecspace, and if Q is a given point, a point R belongs to the space through Q with the given vecspace if and only if the vector of QR is perpendicular to $\mathbf{n}$, that is, if and only if the absolute coordinates of R satisfy the equation

6·41 $$\xi_{\mathbf{n}}(\xi_R-\xi_Q)+\eta_{\mathbf{n}}(\eta_R-\eta_Q)+\zeta_{\mathbf{n}}(\zeta_R-\zeta_Q)+\tau_{\mathbf{n}}(\tau_R-\tau_Q)=0,$$

an equation that is linear but not necessarily homogeneous. Conversely, if we can find any one point Q whose absolute coordinates satisfy a given effective linear equation

6·42 $$\lambda\xi_R+\mu\eta_R+\nu\zeta_R+\varpi\tau_R+\kappa=0,$$

then we shall have

6·43 $$\lambda\xi_Q+\mu\eta_Q+\nu\zeta_Q+\varpi\tau_Q+\kappa=0,$$

and therefore the equation will be equivalent to

6·44 $$\lambda(\xi_R-\xi_Q)+\mu(\eta_R-\eta_Q)+\nu(\zeta_R-\zeta_Q)+\varpi(\tau_R-\tau_Q)=0,$$

which expresses that R belongs to the plane through Q whose vec-

* Always, that is, in real space.

space is perpendicular to the vector $(\lambda, \mu, \nu, \varpi)$, a proper vector since by hypothesis the equation 6·42 is effective; obviously

$$\{-\kappa\lambda/(\lambda^2+\mu^2+\nu^2+\varpi^2),\quad -\kappa\mu/(\lambda^2+\mu^2+\nu^2+\varpi^2),$$
$$-\kappa\nu/(\lambda^2+\mu^2+\nu^2+\varpi^2),\quad -\kappa\varpi/(\lambda^2+\mu^2+\nu^2+\varpi^2)\}$$

is one set* of absolute coordinates satisfying 6·42, and therefore every effective linear equation between the absolute coordinates of a variable point represents one definite space. Two equations can not represent the same space unless one is a mere multiple of the other. The spaces with the equations

$$\lambda'\xi_R+\mu'\eta_R+\nu'\zeta_R+\varpi'\tau_R+\kappa'=0,$$
$$\lambda''\xi_R+\mu''\eta_R+\nu''\zeta_R+\varpi''\tau_R+\kappa''=0$$

are parallel if and only if the vectors $(\lambda', \mu', \nu', \varpi')$, $(\lambda'', \mu'', \nu'', \varpi'')$ are collinear.

A plane has a pair of equations, the equations of a pair of spaces that include it, but there is no one pair of equations that can be regarded as canonical. Any two linear equations taken together represent a plane unless the spaces which they represent individually are parallel.

For a line three equations are required, and three linear equations together represent a line unless two of them represent parallel spaces or represent spaces whose plane of intersection is parallel to the space represented by the third equation.

Ex. xxix. The space through $(3, -1, 5, 1)$ with the vecspace discussed in Example xv (p. 19) has the equation

$$9(\xi-3)+34(\eta+1)+41(\zeta-5)+18(\tau-1)=0,$$

that is,

$$9\xi+34\eta+41\zeta+18\tau-216=0.$$

Ex. xxx. The intersection of a line with a space is calculated most readily if the line is given parametrically and the space by an equation. Thus to find the intersection of the space in the last example with the line through $(-6, -13, 6, 3)$ whose vecline contains $(4, 1, 2, 3)$, we substitute for ξ, η, ζ, τ the values $-6+4x$, $-13+x$, $6+2x$, $3+3x$ typical of the line, and the equation of the space becomes an equation giving the value of x; substituting this value of x, which is 2, we find that the point required is $(2, -11, 10, 9)$.

6·5. Given any point Q and any vecframe **abcd**, any point R may be described by the specification of the vector of QR in the given vecframe. A point Q and a vecframe **abcd** associated for the

* See Appendix, p. 54.

specification of points are said to form a frame of reference. The point Q is called the origin of the frame. A line through Q with vecline containing one of the four vectors of reference is called an axis of the frame, or an axis of reference. A plane of reference is a plane that includes two of the axes, and a space of reference is a space that includes three of them. The words Cartesian and rectangular can be used of the frame if they are applicable to the vecframe.

The coefficients of the vector of $\dot{Q}R$ in the vecframe **abcd** are called the *coordinates* of R in the frame which has Q for origin and **abcd** for vecframe. If these are x_R, y_R, z_R, t_R, the absolute coordinates of R are given by formulae such as

6·51 $$\xi_R = \xi_Q + x_R \xi_{\mathbf{a}} + y_R \xi_{\mathbf{b}} + z_R \xi_{\mathbf{c}} + t_R \xi_{\mathbf{d}}.$$

A substitution of this kind in an equation linear in $\xi_R, \eta_R, \zeta_R, \tau_R$ leads to an equation linear in x_R, y_R, z_R, t_R. Hence in terms of coordinates relative to any frame a space has an equation of exactly the same type as its equation in absolute coordinates.

If the frame is rectangular and Cartesian, a distance l_{RS} from R to S satisfies the equation

6·52 $$l_{RS}^2 = (x_S - x_R)^2 + (y_S - y_R)^2 + (z_S - z_R)^2 + (t_S - t_R)^2,$$

while in terms of absolute coordinates the same distance satisfies the equation

6·53 $$l_{RS}^2 = (\xi_S - \xi_R)^2 + (\eta_S - \eta_R)^2 + (\zeta_S - \zeta_R)^2 + (\tau_S - \tau_R)^2.$$

It follows that no attempt to discover the absolute coordinates of points from a knowledge of relative distances can succeed. The conditions which absolute coordinates must satisfy do not suffice to *determine* these absolute coordinates, since they are fulfilled equally by coordinates relative to any rectangular Cartesian frame of reference, and measurement within a system of points is necessarily ineffective to determine either the absolute origin or the directions of the axes.

Let P be the point whose coordinates in the frame with origin Q and vecframe **abcd** are x_P, y_P, z_P, t_P and let **e**, **f**, **g**, **h** be any four vectors that are not cospatial; then since **e**, **f**, **g**, **h** are expressible in terms of **a**, **b**, **c**, **d** by formulae such as

6·54 $$\mathbf{e} = x_{\mathbf{e}}\mathbf{a} + y_{\mathbf{e}}\mathbf{b} + z_{\mathbf{e}}\mathbf{c} + t_{\mathbf{e}}\mathbf{d},$$

a vector which is given as $X_R\mathbf{e}+Y_R\mathbf{f}+Z_R\mathbf{g}+T_R\mathbf{h}$ can be expressed in terms of **a, b, c, d** by simple substitution, and if this is the vector of a step PR, the vector of the step QR is

$$x_R\mathbf{a}+y_R\mathbf{b}+z_R\mathbf{c}+t_R\mathbf{d}$$

where

6·55 $$x_R = x_P + X_R x_\mathbf{e} + Y_R x_\mathbf{f} + Z_R x_\mathbf{g} + T_R x_\mathbf{h},$$

and so on. That is to say, 6·55 is the first member of a set of formulae expressing the coordinates of any point R in the frame Q**abcd** in terms of the coordinates of the same point in the frame P**efgh**.

The problem of passing from one rectangular Cartesian frame to another is exactly the same as the problem of discovering a single frame of the kind, components relative to the first frame taking the place of absolute components. The origin is arbitrary, and the directions of the axes are given by the formulae of 5·46, interpreted as giving sets of relative components.

7. TRANSLATION, REFLECTION, AND ROTATION

7·1. Two collections of points are said to be *cardinally similar* if it is possible to establish a correlation in which each point in each collection corresponds to one and only one point in the other collection.

If each collection consists of a finite number of points, the collections are cardinally similar if and only if they have the same number of points, but the idea of cardinal similarity is applicable in many cases in which the number of points is not finite. For example, any two planes are cardinally similar, for the typical point in the plane through P' with vecplane $\mathbf{a}'\mathbf{b}'$ is the point R such that the vector of $P'R$ is $x_R'\mathbf{a}'+y_R'\mathbf{b}'$, and the typical point of the plane through P'' with vecplane $\mathbf{a}''\mathbf{b}''$ is the point S such that the vector of $P''S$ is $x_S''\mathbf{a}''+y_S''\mathbf{b}''$; the correlation in which R corresponds to S if and only if x_R' is equal to x_S'' and y_R' to y_S'' is one that establishes the cardinal similarity of the two planes. It will be observed that the correlation in this example is by no means unique: the choice of P'', $\mathbf{a}''$, $\mathbf{b}''$ for the one plane is independent of the choice of P', $\mathbf{a}'$, $\mathbf{b}'$ for the other plane.

A correlation that establishes the cardinal similarity of two sets is called a transformation of one set into the other. If one set is correlated with another and the second with a third, then the first can be correlated directly with the third, and the correlation of the first with the third is said to be compounded of the correlation of the first with the second and the correlation of the second with the third.

A correlation between two collections of points can be regarded simply as a collection of steps. From this point of view, if two correlations can be compounded, and if P_1P_2 is a step belonging to the first correlation and P_2P_3 a step belonging to the second, then P_1P_3 is a step belonging to the compound correlation. It is to be noticed that if a collection of steps is to be a correlation, no two of the steps may have the same beginning and no two of them may have the same end.

The simplest kind of correlation is one in which all the steps are congruent. Such a correlation is called a *translation*, and is specified by the vector common to its steps.

To say that any collection of points can be subjected to any translation is to repeat the assertion of 2·1, that given any point P and any vector $\mathbf{r}$ there is one and only one step which has $\mathbf{r}$ for vector and begins at P, with the addition that if this step is PQ, there is no other step with vector $\mathbf{r}$ which ends at Q.

A correlation may contain any number of zero steps. If a correlation contains the zero step PP, the point P is said to be *fixed* with respect to the transformation defined by the correlation. A simple translation, unless its vector is the zero vector, allows no points to remain fixed. The translation with the zero vector is mere identity, without effect on any point or on any set of points.

If as the result of a transformation the point P comes to Q, the transformation can be compounded of the translation with the vector of PQ and a second correlation. That is to say, any correlation can be compounded of a translation and a correlation with at least one fixed point.

7·2. If the lengths of a step $P'Q'$ are altered by a correlation, that is, if the lengths of $P'Q'$ are different from the lengths of $P''Q''$, where P'' corresponds to P' and Q'' to Q', the step, or any set of points to which P' and Q' belong, is said to be *distorted* by the transformation. A collection of points may be such that no step from one of its members to another is distorted by a particular transformation. A set of points that is not distorted by any one of a number of transformations remains undistorted when these transformations are compounded in any order. It follows from 6·1 that in an undistorted set, angles as well as lengths are unaltered. No set of points is distorted by any translation, for if $P'P''$ and $Q'Q''$ are congruent, having the vector of the translation, so also are $P'Q'$ and $P''Q''$. In this case steps are unaltered in direction as well as in length, but in general the directions of a step *are* liable to change even if the step is from one point to another of an undistorted set.

It is obvious from the general formula deducible from 5·41 to give the measures of a vector in any vecframe, that if two vecframes $\mathbf{a}'\mathbf{b}'\mathbf{c}'\mathbf{d}'$, $\mathbf{a}''\mathbf{b}''\mathbf{c}''\mathbf{d}''$ have the same shape, that is, are such that

7·21 $$a'^2 = a''^2, \quad a'b' \cos \delta_{\mathbf{a}'\mathbf{b}'} = a''b'' \cos \delta_{\mathbf{a}''\mathbf{b}''},$$

and so on, then if frames with any origins Q', Q'' have these vecframes, the correlation in which the point R' whose coordinates in the first frame are x, y, z, t is transformed into the point R'' which has the *same* coordinates x, y, z, t in the *second* frame is a correlation in which no step is distorted.

We must prove a proposition which is to some extent a converse of this. Let two groups of points $Q'A'B'C'D'$, $Q''A''B''C''D''$ be such that corresponding steps in the two groups have the same pairs of lengths and that neither set is in a single space. Then writing $\mathbf{a}'$ for the vector of $Q'A'$, and so on, we have the relations 7·21 satisfied, the first directly on account of the hypothesis, the second in virtue of the trigonometrical identity 6·11. We wish to prove that if two points R', R'' are such that the *five* steps $Q'R'$, $A'R'$, $B'R'$, $C'R'$, $D'R'$ have the same pairs of lengths as the corresponding five steps $Q''R''$, $A''R''$, $B''R''$, $C''R''$, $D''R''$, then the coordinates of R' in the frame with origin Q' and vecframe $\mathbf{a}'\mathbf{b}'\mathbf{c}'\mathbf{d}'$ are the same as the coordinates of R'' in the frame with origin Q'' and vecframe $\mathbf{a}''\mathbf{b}''\mathbf{c}''\mathbf{d}''$. Writing $\mathbf{r}'$, $\mathbf{r}''$ for the vectors of $Q'R'$, $Q''R''$, we have from the conditions to be satisfied, combined with 7·21, four equalities of which the first is

7·22 $$a'r' \cos \delta_{\mathbf{a}'\mathbf{r}'} = a''r'' \cos \delta_{\mathbf{a}''\mathbf{r}''},$$

and expressed in terms of measures and cosines in the frames these take the form

7·23

$$a^2(x'-x'')+ab(y'-y'')\cos\delta_{\mathbf{ab}}+ac(z'-z'')\cos\delta_{\mathbf{ac}}+ad(t'-t'')\cos\delta_{\mathbf{ad}}=0,$$

$$b^2(y'-y'')+ba(x'-x'')\cos\delta_{\mathbf{ba}}+bc(z'-z'')\cos\delta_{\mathbf{bc}}+bd(t'-t'')\cos\delta_{\mathbf{bd}}=0,$$

$$c^2(z'-z'')+ca(x'-x'')\cos\delta_{\mathbf{ca}}+cb(y'-y'')\cos\delta_{\mathbf{cb}}+cd(t'-t'')\cos\delta_{\mathbf{cd}}=0,$$

$$d^2(t'-t'')+da(x'-x'')\cos\delta_{\mathbf{da}}+db(y'-y'')\cos\delta_{\mathbf{db}}+dc(t'-t'')\cos\delta_{\mathbf{dc}}=0,$$

where a is the common value of a' and a'', $\delta_{\mathbf{ab}}$ a common value of $\delta_{\mathbf{a}'\mathbf{b}'}$ and $\delta_{\mathbf{a}''\mathbf{b}''}$, and so on; multiplying these equations by $x'-x''$, $y'-y''$, $z'-z''$, $t'-t''$ and adding, we have an equation that expresses that the vector whose coefficients in one of the vecframes are $x'-x''$, $y'-y''$, $z'-z''$, $t'-t''$ is of zero measure, and since this implies* that the vector is the zero vector, the coefficients vanish separately, which was to be proved.

The only transformations which we propose to discuss are those in which collections of points are undistorted. It follows from what has just been proved that if a set of five points that are not cospatial is undistorted, the transformation is *determined* by its effect on this set of points. If any other point is attached to the set, the point to be attached to the transformed set is known. We can therefore extend the set by supposing any points we wish to belong to it, attaching the corresponding points to the transformed set. In particular, if we have any point whose pairs of distances from five points that are not cospatial are the same as its pairs of distances from the five corresponding points, we can attach this point to both sets and regard it as fixed with respect to the transformation.

7·3. It is possible for all the points in a space to be fixed with respect to a transformation in which a collection of points remains undistorted. To examine the nature of the transformation when this occurs, let rectangular Cartesian axes be taken with the origin in the space and with the first axis perpendicular to the space. If then the point R', with coordinates (x', y', z', t'), is transformed to R'', with coordinates (x'', y'', z'', t''),

7·31 $$x'^2+(y'-y_Q)^2+(z'-z_Q)^2+(t'-t_Q)^2$$
$$=x''^2+(y''-y_Q)^2+(z''-z_Q)^2+(t''-t_Q)^2$$

* Not in complex space; see Appendix, p. 55.

for *every* position of Q in the space for which x is zero; taking y_Q, z_Q, t_Q all zero, we have

7·32 $$x'^2 + y'^2 + z'^2 + t'^2 = x''^2 + y''^2 + z''^2 + t''^2,$$

and on account of this equality 7·31 becomes

7·33 $$(y' - y'')\, y_Q + (z' - z'')\, z_Q + (t' - t'')\, t_Q = 0,$$

implying

7·34 $$y' = y'', \quad z' = z'', \quad t' = t'',$$

and reducing 7·32 to

7·35 $$x'^2 = x''^2.$$

Since a step between points that correspond is to be undistorted even if neither point is in the fixed plane, we are to have

7·36 $$(x_1' - x_2')^2 + (y_1' - y_2')^2 + (z_1' - z_2')^2 + (t_1' - t_2')^2$$
$$= (x_1'' - x_2'')^2 + (y_1'' - y_2'')^2 + (z_1'' - z_2'')^2 + (t_1'' - t_2'')^2$$

whenever R_1' corresponds to R_1'' and R_2' to R_2''. Making use of the relations already found, namely

7·37 $$y_1' = y_1'', \quad z_1' = z_1'', \quad t_1' = t_1'',$$
$$y_2' = y_2'', \quad z_2' = z_2'', \quad t_2' = t_2'',$$
$$x_1'^2 = x_1''^2, \quad x_2'^2 = x_2''^2,$$

we have from 7·36,

7·38 $$x_1' x_2' = x_1'' x_2'',$$

and this is the only additional relation necessary. Thus if x_1' coincides with x_1'' then x_2' coincides with x_2'', and therefore on account of 7·37, R_2' coincides with R_2'': the transformation is a mere identity, correlating a set with itself. But if x_1' is equal to $-x_1''$, then x_2' is equal to $-x_2''$, and the transformation is expressed by

7·39 $$x' + x'' = 0, \quad y' = y'', \quad z' = z'', \quad t' = t''.$$

The transformation in the latter case is characterised by the property that the line joining any two points that correspond is perpendicular to a fixed space and is bisected by that space; the transformation is called a *reflection* in the fixed space. What we have proved is that any transformation which leaves a collection of points undistorted and leaves all the points in a particular space fixed, either leaves the collection wholly unaltered or reflects the collection in the fixed space.

7·4. The absolute coordinates of a point R that is equidistant from two given points P, Q satisfy the equation

$$(\xi_R-\xi_Q)^2+(\eta_R-\eta_Q)^2+(\zeta_R-\zeta_Q)^2+(\tau_R-\tau_Q)^2$$
$$=(\xi_R-\xi_P)^2+(\eta_R-\eta_P)^2+(\zeta_R-\zeta_P)^2+(\tau_R-\tau_P)^2,$$

which is equivalent to

7·41 $$(\xi_Q-\xi_P)\{\xi_R-\tfrac{1}{2}(\xi_P+\xi_Q)\}+(\eta_Q-\eta_P)\{\eta_R-\tfrac{1}{2}(\eta_P+\eta_Q)\}$$
$$+(\zeta_Q-\zeta_P)\{\zeta_R-\tfrac{1}{2}(\zeta_P+\zeta_Q)\}+(\tau_Q-\tau_P)\{\tau_R-\tfrac{1}{2}(\tau_P+\tau_Q)\}=0,$$

and therefore expresses that R lies in a definite space, which is perpendicular to the step PQ and contains the midpoint of PQ.

It follows that if S', S'' are distinct corresponding points in a correlation that leaves a set of points undistorted and leaves a particular point P fixed, the space that is the perpendicular bisector of $S'S''$ is a space that contains P and contains also all the other points that are fixed. If all the points of this space are fixed, then since S' does not coincide with S'' the transformation is a reflection in the space. If this is not the case, and if T' is a point of the space that does not coincide with the corresponding point T'', the perpendicular bisectors of $S'S''$ and $T'T''$ do not coincide, since the former contains R' and the latter does not*, nor are these spaces parallel but distinct, since the point P is common to them. Hence the points common to the two perpendicular bisectors compose a definite plane p, which contains the point P but does not contain any of the points S', S'', T', T''. If Q, R are any two points of p not collinear with P, then because Q, R belong to p, not only have the steps PQ, PR, PS', PT', QR, $S'T'$ the same lengths as the steps PQ, PR, PS'', PT'', QR, $S''T''$, but the steps QS', QT', RS', RT' have the same lengths as the steps QS'', QT'', RS'', RT'', and because Q, R are not collinear with P, the five points P, Q, R, S', T' are not cospatial. It follows that the transformation is definable as *the* transformation in which $PQRS'T'$ becomes $PQRS''T''$ and no distortion takes place. The effect of the transformation on any point U' is to be found by attaching U' to the set $PQRS'T'$; the correlative of U'' is attached to $PQRS''T''$ in the same way, that is, by steps of the same length as those from P, Q, R, S', T' to U'. In other words, if $\mathbf{a}$, $\mathbf{b}$, $\mathbf{c}'$, $\mathbf{d}'$, $\mathbf{c}''$, $\mathbf{d}''$ are the

* The argument is not valid for complex space, and the whole notion of rotation is complicated if self-perpendicular elements are admitted.

vectors of PQ, PR, PS', PT', PS'', PT'', the point reached from P by a step with any vector $x\mathbf{a}+y\mathbf{b}+z\mathbf{c}'+t\mathbf{d}'$ corresponds to the point reached from P by the step whose vector is $x\mathbf{a}+y\mathbf{b}+z\mathbf{c}''+t\mathbf{d}''$. In particular, if z and t are zero, the step from P is unchanged: every point in the plane p is a fixed point. Also a point in the space $PQRS'$ is a point for which t is zero and therefore corresponds to a point in the space $PQRS''$: if any number of points are in one space which includes the plane p, their correlatives also are in one space which includes this plane. If the step joining two points U', V' has vector $(x_V-x_U)\,\mathbf{a}+(y_V-y_U)\,\mathbf{b}+(z_V-z_U)\,\mathbf{c}'+(t_V-t_U)\,\mathbf{d}'$, the step joining the corresponding points U'', V'' has the vector $(x_V-x_U)\,\mathbf{a}+(y_V-y_U)\,\mathbf{b}+(z_V-z_U)\,\mathbf{c}''+(t_V-t_U)\,\mathbf{d}''$; the step $U'V'$ is parallel to the plane p only if z_V-z_U and t_V-t_U are both zero, and if these conditions are satisfied $U''V''$ is congruent with $U'V'$.

It is now easy to describe the transformation in detail. The space that bisects perpendicularly any proper step in the correlation necessarily contains every fixed point and therefore includes the plane p. Hence every step is perpendicular to p, and if U' is any point of the set to be transformed, the correlative U'' is in the plane u through U' perpendicular to p. If the plane through U' parallel to p cuts a plane w parallel to u in W', the correlative of W' is determinable as the point in which the plane through U'' parallel to p cuts w. The step $U'W'$ is congruent with $U''W''$, and therefore the step $W'W''$ is congruent with $U'U''$. Thus the transformation in one plane perpendicular to p copies the transformation in any other plane perpendicular to p.

To discover the features of the transformation in one plane q perpendicular to p, we may use at last not only the language but also the propositions of elementary geometry. Let $\mathbf{c}$, $\mathbf{d}$ be perpendicular unit vectors in the vecplane of q, and let Q be the point in which p cuts q. In any ordinary plane q_* take a pair of axes O_*X_*, O_*Y_* at right angles, and associate with the point R of q which is such that the vector of QR is $z_R\mathbf{c}+t_R\mathbf{d}$, the point of q_* whose coordinates with respect to O_*X_* and O_*Y_* are z_R and t_R; denote this ordinary point by R_*. Comparison of formulae shews that the lengths of two steps ST, UV in q are equal if and only if the lengths of the corresponding steps S_*T_*, U_*V_* in q_* are equal. The transformation in q which leaves all steps undistorted and leaves Q

fixed therefore corresponds to a transformation in q_* which leaves all steps undistorted and leaves O_* fixed, and such a transformation is either a reflection in some line through O_* in the plane or a rotation about O_* through some constant angle. A reflection in a line through O_* has all the points of that line for fixed points and can not correspond to a transformation in q in which there is no point fixed except Q. Hence the transformation in q_* is necessarily a rotation about O_*, and in q measured steps retain their lengths while their directions change through angles that are equal if measured conformably. It follows that the whole transformation can be described thus:

There is a plane p which consists entirely of fixed points; if U', V' are any two points not in this plane, if Q, R are the points in which p cuts the planes through U', V' perpendicular to p, and if U'', V'' are the correlatives of U', V', then QU'', RV'' have the same pairs of lengths as QU', RV', the directions in which QU' and QU'' have one common length and the directions in which RV' and RV'' have one common length are all coplanar, being perpendicular to p, and the angles from the first of these directions to the second are the same as the angles from the third to the fourth provided that these angles are measured conformably.

Naturally the transformation is described as a *rotation* about the fixed plane p through a definite angle. There is difficulty in describing the rotation by means of the plane and the angle alone, since it is necessary to distinguish between rotation through an angle θ and rotation through the angle $-\theta$. We have evaded the difficulty above by comparing the effects of the transformation on two points instead of describing its effect on a single point. The alternative plan is to specify two directions $\mathbf{s}$, $\mathbf{t}$ perpendicular to p and an angle $\epsilon_{\mathbf{st}}$ from the first of them to the second. We can then say that there is an angle θ such that if U' is any point not in p and if Q is the point in which the plane through U' perpendicular to p is cut by p, the steps QU', QU'' have the same pair of lengths, and the directions in which they have one of their common lengths are coplanar with $\mathbf{s}$ and $\mathbf{t}$ and are such that θ is an angle from the first of them to the second conformable with $\epsilon_{\mathbf{st}}$.

It is a simple matter to find a set of formulae representing a given rotation. Let $\mathbf{a}$, $\mathbf{b}$ be perpendicular unit vectors in the vecplane of the fixed plane, and let **c**, **d** be perpendicular unit vectors in the vecplane perpendicular to that of the fixed plane; the assumption that $\frac{1}{2}\pi$ is to be an angle *from* **c** *to* **d** supplies a standard for the conformable measurement of angles, and we can speak of the angle of rotation θ. Let the origin be a point Q in the fixed plane, and denote by C, D the points such that the vectors of QC, QD are the unit vectors **c**, **d**. If by the rotation C, D are made to correspond to E, F, the vectors of QE, QF, which we will denote by **e**, **f**, are unit vectors, and θ is an angle from **c** to **e** and from **d** to **f** conformable with the angle $\frac{1}{2}\pi$ from **c** to **d** and from **e** to **f**. By the definitions of sines and cosines, we have

$$\cos \epsilon_{\mathbf{ce}} = x_{\mathbf{c}} x_{\mathbf{e}} + y_{\mathbf{c}} y_{\mathbf{e}}, \quad \sin \epsilon_{\mathbf{ce}} = x_{\mathbf{c}} y_{\mathbf{e}} - y_{\mathbf{c}} x_{\mathbf{e}},$$

$$\cos \epsilon_{\mathbf{df}} = x_{\mathbf{d}} x_{\mathbf{f}} + y_{\mathbf{d}} y_{\mathbf{f}}, \quad \sin \epsilon_{\mathbf{df}} = x_{\mathbf{d}} y_{\mathbf{f}} - y_{\mathbf{d}} x_{\mathbf{f}},$$

where $x_{\mathbf{r}}\mathbf{c} + y_{\mathbf{r}}\mathbf{d}$ denotes temporarily the typical expression of a vector **r** coplanar with **c** and **d**. Utilising the obvious values of $x_{\mathbf{c}}$, $y_{\mathbf{c}}$, $x_{\mathbf{d}}$, $y_{\mathbf{d}}$, we have

7·42 $\quad x_{\mathbf{e}} = \cos\theta, \quad y_{\mathbf{e}} = \sin\theta, \quad x_{\mathbf{f}} = -\sin\theta, \quad y_{\mathbf{f}} = \cos\theta,$

and therefore

7·43 $\quad \mathbf{e} = \mathbf{c}\cos\theta + \mathbf{d}\sin\theta, \quad \mathbf{f} = -\mathbf{c}\sin\theta + \mathbf{d}\cos\theta.$

But we have already seen that the point R'' corresponds to the point R' if the vector of PR' is $x'\mathbf{a} + y'\mathbf{b} + z'\mathbf{c} + t'\mathbf{d}$ and that of PR'' is $x'\mathbf{a} + y'\mathbf{b} + z'\mathbf{e} + t'\mathbf{f}$; substituting for **e** and **f** from 7·43, we have for the vector of PR'' the expression

$$x'\mathbf{a} + y'\mathbf{b} + (z'\cos\theta - t'\sin\theta)\,\mathbf{c} + (z'\sin\theta + t'\cos\theta)\,\mathbf{d}.$$

That is to say, the transformation is given by

7·44 $\quad x'' = x', \; y'' = y', \; z'' = z'\cos\theta - t'\sin\theta, \; t'' = z'\sin\theta + t'\cos\theta,$

provided that the first two axes are perpendicular lines in the fixed plane and the last two are perpendicular lines perpendicular to that plane. If we replace $\cos\theta$, $\sin\theta$ by their values in terms of $\tan\frac{1}{2}\theta$ and write k for this tangent, the formulae of the transformation become

7·45 $$x'' = x', \quad y'' = y', \quad (1 + k^2)\, z'' = (1 - k^2)\, z' - 2kt',$$
$$(1 + k^2)\, t'' = 2kz' + (1 - k^2)\, t'.$$

7·5. A reflection is distinguished from a rotation by the existence of a fixed space, but this is not the fundamental distinction between the two types of transformation, for on the compounding of the transformation with a translation the fixed points, whether they compose a plane or a space, may disappear altogether. A difference of another kind can be explained by means of the set of formulae 7·39.

It follows from the possibility of adding coplanar angles, that if a rotation through an angle ϕ about a plane p is compounded with a rotation through an angle ψ about the same plane, the result is a rotation through the angle $\phi+\psi$ about that plane. Hence in considering a rotation through an angle α about a plane p, we may regard α as the final value of a variable θ which takes a succession of values from 0 to α, and we may regard the rotation as the result of compounding a succession of rotations through angles whose sum is α. If we place no restriction on the number of the component rotations, we may make the angles of these components as small as we please, and therefore we may keep the lengths of the step which any *given* point takes in any single component rotation below any assigned limit. The rotation through the angle α may therefore be developed as a *continuous* transformation *throughout* which there is no distortion; the fixed plane is fixed not merely *with respect to* the resultant transformation but *throughout* the continuous transformation.

The case of a reflection is altogether different. There are of course many ways in which we can regard the transformation

7·51 $$x''=-x',\quad y''=y',\quad z''=z',\quad t''=t'$$

as derived from a succession of transformations. For example, we may replace 7·51 by

7·52 $$x''=kx',\quad y''=y',\quad z''=z',\quad t''=t'$$

and regard k as varying between 1 and -1. But with this choice, k can not vary *from* 1 *to* -1 since the value 0 is impossible for k; also not more than one of the intermediate transformations can be a reflection, and therefore, since every one of these transformations has a fixed space, there is distortion in the intermediate stages. Nothing could be more rash than to assume either from the failure of the most obvious succession of transformations or from the nature of reflection in ordinary space that no presentation of a reflection as the result of a continuous transformation admitting undistorted sets is possible, but in fact the conclusion can be reached on sure ground* and we must base further definitions upon it.

A transformation with respect to which all collections of points are undistorted is described as a *displacement* or a *perversion* according as it can or can not be represented as the result of a continuous transformation throughout which there is no distortion. It is obvious that if two displacements are compounded the result is a displacement, and it follows that if a displacement and a perversion are compounded the result is a perversion. Hence if any finite number of displacements and perversions are compounded, the result is a displacement or a perversion according as the number of perversions is even or odd.

We have shewn that every rotation is a displacement, and we shall assume that every reflection is a perversion. Every translation is a displacement, since the translation whose vector is $\mathbf{r}$ may be regarded as the resultant of n translations each with vector $\mathbf{r}/n$. It follows that every displacement is expressible as the resultant of a translation and a rotation, and every perversion as the resultant of a translation and a reflection; but the analysis of a given displacement or perversion into a translation and a rotation or reflection is not unique except in the trivial case in which the transformation is a pure translation.

7·6. To justify the use of geometrical terms in algebra, we need only compare the set of formulae 5·46 with the theorem that a transformation in which they are employed is either a reflection or a rotation. There is nothing in 5·46, or in fact in Cayley's theory of such transformations, to suggest the result.

But ultimately the results are algebraic. We can assert for example the following theorem:

If two sets of variables $(\xi', \eta', \zeta', \tau')$ and $(\xi'', \eta'', \zeta'', \tau'')$ are connected in such a way that whenever $(\xi_1', \eta_1', \zeta_1', \tau_1')$ and $(\xi_2', \eta_2', \zeta_2', \tau_2')$ correspond to $(\xi_1'', \eta_1'', \zeta_1'', \tau_1'')$ and $(\xi_2'', \eta_2'', \zeta_2'', \tau_2'')$ the relation

7·61
$$(\xi_2' - \xi_1')^2 + (\eta_2' - \eta_1')^2 + (\zeta_2' - \zeta_1')^2 + (\tau_2' - \tau_1')^2$$
$$= (\xi_2'' - \xi_1'')^2 + (\eta_2'' - \eta_1'')^2 + (\zeta_2'' - \zeta_1'')^2 + (\tau_2'' - \tau_1'')^2$$

* The determinant formed of the coefficients of the vectors in 5·46 has one of the values $+1$, -1 whatever the values of the six parameters; a *continuous* change in the parameters can not effect a change from one of the values of the determinant to the other, which can be caused only by the reversal of one or of three of the arbitrary signs.

is satisfied, then either it is possible to find a substitution of the form

7·62

$$\begin{aligned}\xi' &= X' + \lambda_1 x' + \lambda_2 y' + \lambda_3 z' + \lambda_4 t', & \xi'' &= X'' + \lambda_1 x'' + \lambda_2 y'' + \lambda_3 z'' + \lambda_4 t'',\\ \eta' &= Y' + \mu_1 x' + \mu_2 y' + \mu_3 z' + \mu_4 t', & \eta'' &= Y'' + \mu_1 x'' + \mu_2 y'' + \mu_3 z'' + \mu_4 t'',\\ \zeta' &= Z' + \nu_1 x' + \nu_2 y' + \nu_3 z' + \nu_4 t', & \zeta'' &= Z'' + \nu_1 x'' + \nu_2 y'' + \nu_3 z'' + \nu_4 t'',\\ \tau' &= T' + \varpi_1 x' + \varpi_2 y' + \varpi_3 z' + \varpi_4 t', & \tau'' &= T'' + \varpi_1 x'' + \varpi_2 y'' + \varpi_3 z'' + \varpi_4 t'',\end{aligned}$$

in which the coefficients satisfy the conditions* that ensure the identities

7·63
$$(\xi_2' - \xi_1')^2 + (\eta_2' - \eta_1')^2 + (\zeta_2' - \zeta_1')^2 + (\tau_2' - \tau_1')^2 = (x_2' - x_1')^2 + (y_2' - y_1')^2 + (z_2' - z_1')^2 + (t_2' - t_1')^2,$$
$$(\xi_2'' - \xi_1'')^2 + (\eta_2'' - \eta_1'')^2 + (\zeta_2'' - \zeta_1'')^2 + (\tau_2'' - \tau_1'')^2 = (x_2'' - x_1'')^2 + (y_2'' - y_1'')^2 + (z_2'' - z_1'')^2 + (t_2'' - t_1'')^2,$$

which reduces the relation between the sets of variables to

7·64 $$x' + x'' = 0, \quad y' = y'', \quad z' = z'', \quad t' = t'',$$

or it is possible to find a substitution of the same form with the same relations between the coefficients which reduces the relation to

7·65 $$x'' = x', \quad y'' = y', \quad (1 + k^2) z'' = (1 - k^2) z' - 2kt',$$
$$(1 + k^2) t'' = 2kz' + (1 - k^2) t',$$

where k is some definite constant; in the latter case it is possible also to find a substitution

7·66

$$\xi' = X'(s) + \lambda_1(s) x'(s) + \lambda_2(s) y'(s) + \lambda_3(s) z'(s) + \lambda_4(s) t'(s),$$
$$\xi'' = X''(s) + \lambda_1(s) x''(s) + \lambda_2(s) y''(s) + \lambda_3(s) z''(s) + \lambda_4(s) t''(s),$$

and so on, in which the twenty-four coefficients are continuous functions of a single variable s whose values satisfy the imposed conditions for every value of s, and which is such that for one particular value of s the variables $x'(s)$, $x''(s)$, ... are the variables ξ', ξ'', ... and for another value of s these variables reduce to x', x'', ...; in the former case no such continuous substitution exists.

* These are the four conditions of the form

$$\lambda_r^2 + \mu_r^2 + \nu_r^2 + \varpi_r^2 = 1,$$

and the six comprised in

$$\lambda_r \lambda_s + \mu_r \mu_s + \nu_r \nu_s + \varpi_r \varpi_s = 0, \qquad r \neq s.$$

If given a constant K we alter certain of the variables by writing $K\tau'$, $K\tau''$, Kt', Kt'' for τ', τ'', t', t'' and certain of the coefficients by writing λ_4/K, μ_4/K, ν_4/K for λ_4, μ_4, ν_4, and KT', KT'', $K\varpi_1$, $K\varpi_2$, $K\varpi_3$ for T', T'', ϖ_1, ϖ_2, ϖ_3, the formulae of substitution are unaltered, but 7·61 becomes

$$(\xi_2'-\xi_1')^2+(\eta_2'-\eta_1')^2+(\zeta_2'-\zeta_1')^2+K^2(\tau_2'-\tau_1')^2$$
$$=(\xi_2''-\xi_1'')^2+(\eta_2''-\eta_1'')^2+(\zeta_2''-\zeta_1'')^2+K^2(\tau_2''-\tau_1'')^2,$$

the identities 7·63 take corresponding forms, and the conditions satisfied by the coefficients are

$$\lambda_r^2+\mu_r^2+\nu_r^2+K^2\varpi_r^2=1,$$
$$\lambda_r\lambda_s+\mu_r\mu_s+\nu_r\nu_s+K^2\varpi_r\varpi_s=0, \qquad r\neq s.$$

Supposing the transformation to be a displacement, the substitution of w/K for k replaces 7·65 by

$$x''=x', \quad y''=y', \quad \{1+(w^2/K^2)\}\,z''=\{1-(w^2/K^2)\}\,z'-2wt',$$
$$\{1+(w^2/K^2)\}\,t''=2(w/K^2)\,z'+\{1-(w^2/K^2)\}\,t'.$$

It is only in its square that K now enters the formulae, and replacing K^2 by L, we find that if the identity

7·67 $$(\xi_2'-\xi_1')^2+(\eta_2'-\eta_1')^2+(\zeta_2'-\zeta_1')^2+L(\tau_2'-\tau_1')^2$$
$$=(\xi_2''-\xi_1'')^2+(\eta_2''-\eta_1'')^2+(\zeta_2''-\zeta_1'')^2+L(\tau_2''-\tau_1'')^2$$

is maintained throughout a continuous transformation, the transformation is reducible to the form

7·68 $$x''=x', \quad y''=y', \quad \{1+(w^2/L)\}\,z''=\{1-(w^2/L)\}\,z'-2wt',$$
$$\{1+(w^2/L)\}\,t''=2(w/L)\,z'+\{1-(w^2/L)\}\,t',$$

by linear substitutions of the form in 7·62. This is a theorem in pure algebra, admitting theoretically of direct verification, and we have little hesitation in affirming that its truth is not dependent on the sign of L, and therefore that if the identity maintained is

$$(\xi_2'-\xi_1')^2+(\eta_2'-\eta_1')^2+(\zeta_2'-\zeta_1')^2-C^2(\tau_2'-\tau_1')^2$$
$$=(\xi_2''-\xi_1'')^2+(\eta_2''-\eta_1'')^2+(\zeta_2''-\zeta_1'')^2-C^2(\tau_2''-\tau_1'')^2,$$

then the transformation is reducible to

7·69 $$x''=x', \quad y''=y', \quad \{1-(w^2/C^2)\}\,z''=\{1+(w^2/C^2)\}\,z'-2wt',$$
$$\{1-(w^2/C^2)\}\,t''=2(w/C^2)\,z'+\{1+(w^2/C^2)\}\,t'.$$

The connection of this analysis with the theory of relativity is easily seen. It is an induction from the most careful experiments that the velocity of light has a common value C for all observers, and it is not difficult to deduce that it is impossible to compare distances in space at different times. The only comparison possible is between events, and the only measure on which different observers necessarily agree is a measure known as the *interval* between two events. The interval is a function dependent on the relative positions of the events in space and time, and if an absolute framework fixed in time and space could be established the square of the interval between two events would be the difference between the square of their separation in space and the product by C^2 of the square of their separation in time, that is to say, would have the form

$$(\xi_2 - \xi_1)^2 + (\eta_2 - \eta_1)^2 + (\zeta_2 - \zeta_1)^2 - C^2 (\tau_2 - \tau_1)^2.$$

It follows firstly that no measuring of intervals *can* disclose an absolute framework, since the conditions to be satisfied by absolute coordinates do not differ from those to be satisfied by relative coordinates. It follows also that if two observers do not agree in the measurement of time, the relation between their measures of time and space is necessarily reducible to the form given in 7·69.

APPENDIX

·1. A mathematician can construct a space as readily with complex numbers as with real numbers. A point in a four-dimensional complex space depends ultimately on eight real numbers, but the geometry of this space differs from that of eight-dimensional real space not only in the relative importance of problems that the two geometries have in common but in the very definitions of distance and perpendicularity.

The fundamental distinction between real space and complex space is that in the latter there are proper self-perpendicular elements. The difficulty of ascribing *direction* to a nul line can be overcome by the Frege-Russell method, but other problems remain: for example, since a nul line is parallel to any space to which it is perpendicular, the definition given in 6·3 of the distances between a point and a space may fail. To deal adequately with complex space would have been to change the whole scale of the booklet and to offer it to a far more limited circle of readers. Paragraphs ·2, ·3, ·4, ·6 of this Appendix deal only with details in the demonstrations of certain propositions true of complex space but proved in the text by methods available for real space alone.

It is commonly *said*, in work on three-dimensional space, that a vector has one definite measure, and that this is essentially positive unless it is zero. But there is no doubt that the duplex vector is a *practical* necessity in real three-dimensional space and a *theoretical* necessity in complex space of any order. In applied mathematics we never hesitate to treat a force whose line of action is known as a force of amount R in a particular direction along the line; if on calculation R proves to be negative, we say that the direction must 'really' be reversed, and we never dream of questioning the deduction on the ground that a vector with negative measure has not even been defined. With complex numbers, there is no classification of square roots that resembles* the grouping of square roots

* In the case of a complex number $x+iy$, we can of course distinguish unless y is zero and x is negative between the square root whose real part is positive and the square root whose real part is negative, but since the product of two square roots with real part positive does not necessarily have its own real part positive, an arbitrary distinction of this kind is of no practical value.

of real numbers as positive or negative; it is impossible to progress without recognising two measures to every proper vector, and unless the duplex vector has been familiar in real space, the whole vocabulary of the subject changes and one of the principal advantages of using geometrical terms in discussing sets of numbers disappears.

·2. See p. 18.

The conclusion

·21 $$\lambda a+\mu b+\nu c+\varpi d=0$$

can always be avoided if $\lambda, \mu, \nu, \varpi$ are not all zero, even if the numbers are complex. If λ is not zero, we may take b, c, d zero and a not zero.

·3. See p. 18.

To establish the uniqueness of the vecline perpendicular to a given vecspace **abc** without the assumption that a vecline can not be self-perpendicular, suppose that there are two sets of numbers $(\lambda', \mu', \nu', \varpi')$ and $(\lambda'', \mu'', \nu'', \varpi'')$ each of which satisfies 4·24 and that the set of ratios $\lambda':\mu':\nu':\varpi'$ is different in some respect from the set of ratios $\lambda'':\mu'':\nu'':\varpi''$. Then there is no real loss of generality in supposing that $\lambda'\mu''-\mu'\lambda''$ is not zero, and the pair of equations

·31 $$\lambda'\xi+\mu'\eta+\nu'\zeta+\varpi'\tau=0, \quad \lambda''\xi+\mu''\eta+\nu''\zeta+\varpi''\tau=0$$

is equivalent to the pair

·32 $$\xi=k\zeta+l\tau, \quad \eta=m\zeta+n\tau,$$

whatever the values of ξ, η, ζ, τ, where

·33 $$k=\frac{\mu'\nu''-\nu'\mu''}{\lambda'\mu''-\mu'\lambda''}, \quad l=\frac{\mu'\varpi''-\varpi'\mu''}{\lambda'\mu''-\mu'\lambda''},$$
$$m=\frac{\nu'\lambda''-\lambda'\nu''}{\lambda'\mu''-\mu'\lambda''}, \quad n=\frac{\varpi'\lambda''-\lambda'\varpi''}{\lambda'\mu''-\mu'\lambda''}.$$

Thus

·34 $$\xi_{\mathbf{a}}=k\zeta_{\mathbf{a}}+l\tau_{\mathbf{a}}, \quad \xi_{\mathbf{b}}=k\zeta_{\mathbf{b}}+l\tau_{\mathbf{b}}, \quad \xi_{\mathbf{c}}=k\zeta_{\mathbf{c}}+l\tau_{\mathbf{c}},$$

·35 $$\eta_{\mathbf{a}}=m\zeta_{\mathbf{a}}+n\tau_{\mathbf{a}}, \quad \eta_{\mathbf{b}}=m\zeta_{\mathbf{b}}+n\tau_{\mathbf{b}}, \quad \eta_{\mathbf{c}}=m\zeta_{\mathbf{c}}+n\tau_{\mathbf{c}}.$$

But by the general algebraic theorem enunciated in 4·2, there is at least one effective linear relation between the three dyads $(\zeta_{\mathbf{a}}, \tau_{\mathbf{a}})$, $(\zeta_{\mathbf{b}}, \tau_{\mathbf{b}})$, $(\zeta_{\mathbf{c}}, \tau_{\mathbf{c}})$, that is, at least one set of numbers (a, b, c) not all zero such that simultaneously

·36 $$a\zeta_{\mathbf{a}}+b\zeta_{\mathbf{b}}+c\zeta_{\mathbf{c}}=0, \quad a\tau_{\mathbf{a}}+b\tau_{\mathbf{b}}+c\tau_{\mathbf{c}}=0,$$

and it follows from ·34 and ·35 that with these values of a, b, c

·37 $$a\xi_{\mathbf{a}} + b\xi_{\mathbf{b}} + c\xi_{\mathbf{c}} = 0, \quad a\eta_{\mathbf{a}} + b\eta_{\mathbf{b}} + c\eta_{\mathbf{c}} = 0.$$

Taken together, ·37 and ·36 are equivalent to the equation

·38 $$a\mathbf{a} + b\mathbf{b} + c\mathbf{c} = \mathbf{0},$$

and this equation contradicts the hypothesis that **a**, **b**, **c** are not coplanar.

·4. See p. 36.

In complex space the denominator might be zero. But if λ is not zero, the point $(-\kappa/\lambda, 0, 0, 0)$ is in the space, and by hypothesis at least one of the four coefficients λ, μ, ν, ϖ is different from zero. The point used in the text is the point in which the space is met by the line through the origin perpendicular to it, and there is no longer such a point available if the line is *parallel as well as perpendicular* to the space.

·5. It is no more difficult to construct a space of any finite number n of dimensions than to build one of four dimensions; the point is a set of numbers $(\xi^{(1)}, \xi^{(2)}, \ldots, \xi^{(n)})$. There are two methods of forming subregions of space.

The limiting of R by a parametric set of equations

$$\xi_R{}^{(r)} = \xi_Q{}^{(r)} + x_R{}^{(1)}\,\xi_1{}^{(r)} + x_R{}^{(2)}\,\xi_2{}^{(r)} + \ldots + x_R{}^{(k)}\,\xi_k{}^{(r)}, \quad r = 1, 2, \ldots, n,$$

in which $(\xi_h{}^{(1)}, \xi_h{}^{(2)}, \ldots, \xi_h{}^{(n)})$, for all values of h from 1 to k, is a definite set of numbers and $x_R{}^{(1)}, x_R{}^{(2)}, \ldots, x_R{}^{(k)}$ take all possible values, corresponds to the method in the text, the simplest form of region of this kind being the line, for which k is unity. Alternatively R may be restricted by a set of equations

$$\lambda_1{}^{(p)}\,\xi_R{}^{(1)} + \lambda_2{}^{(p)}\,\xi_R{}^{(2)} + \ldots + \lambda_n{}^{(p)}\,\xi_R{}^{(n)} + \kappa^{(p)} = 0, \quad p = 1, 2, \ldots, q,$$

in which $(\lambda_1{}^{(p)}, \lambda_2{}^{(p)}, \ldots, \lambda_n{}^{(p)})$ are given sets of coefficients; the simplest restriction of this kind is by a single equation, and it is usual to use the familiar name of plane for this subregion, which is of $n-1$ dimensions, not for the subregion which follows the line in the parametric order.

I have broken with convention in the text, firstly to avoid repelling the general reader by coined or formal words, and secondly because it is the essence of Einstein's and Minkowski's idea that a *plane section* of the space-time continuum *is* space in the ordinary sense. Had my object been different, tampering with the accepted vocabulary would have been unpardonable, and had I been dealing with more dimensions than four, useless, but with any other purpose in view, I should have been addressing myself exclusively to mathematicians and the restriction to four dimensions would have been absurd.

·6. See p. 41.

The condition for the set of equations 7·23 to be satisfied by a set of numbers $x'-x''$, $y'-y''$, $z'-z''$, $t'-t''$ not all zero, is identical with the condition for the expression for $rs \cos \delta_{\mathbf{rs}}$ to be a product of linear factors. In complex space, this is not an impossible condition, but it is equivalent to the assumption that the space of four dimensions is an isotropic section of space of five dimensions, and entails properties of which the failure of the result in which 7·23 was used is among the least bizarre.

·7. It should be added that only the *special* principle of relativity has its mathematical foundation in the elementary work which we have been doing. The *general* principle of relativity that is invoked in the discussion of gravitation requires mathematics far more elaborate. But the four-dimensional continuum whose properties are used is again *constructed by the mathematician*, and the study of it is as free from *psychological* difficulty as is that of the simple space which we have considered.

INDEX OF TERMS DEFINED

Printed in the United States
By Bookmasters